Mastering Mathcad
Version 7

Howard Keller

John G. Crandall

McGraw-Hill

New York San Francisco Washington, D.C. Auckland Bogotá
Caracas Lisbon London Madrid Mexico City Milan
Montreal New Delhi San Juan Singapore
Sydney Tokyo Toronto

Library of Congress Cataloging-in-Publication Data

Keller, Howard, 1941–
    Mastering Mathcad version 7 / Howard Keller, John G. Crandall.
        p.   cm.
    Includes index.
        ISBN 0-07-913136-0 (book/CD-ROM set). – ISBN 0-07-034310-1 (book
    only). – ISBN 0-07-853037-7 (CD-ROM only)
        1. MathCAD.   2. Mathematics–Data processing.   I. Crandall, John
    G. (John Gaylord)   II. Title.
    QA76.95.K45   1998
    510′.285′5369–dc21

                                                            97-51231
                                                            CIP

McGraw-Hill

A Division of The McGraw·Hill Companies

1 2 3 4 5 6 7 8 9 0   DOC/DOC   9 0 3 2 1 0 9 8

ISBN 0-07-034310-1
part of
0-07-913136-0

The sponsoring editor for this book was Steve Chapman, the editing supervisor was Bernard Onken, and the production supervisor was Clare Stanley. It was set in Century Schoolbook by ATLIS Graphics and Design.

Printed and bound by R.R. Donnelley & Sons Company.

McGraw-Hill books are available at special quantity discounts to use as premiums and sales promotions, or for use in corporate training programs. For more information, please write to the Director of Special Sales, McGraw-Hill, 11 West 19th Street, New York, NY 10011. Or contact your local bookstore.

This book is printed on recycled, acid-free paper containing a minimum of 50% recycled de-inked fiber.

Contents

Preface

If, like the authors of this book, you went to college in the days before the personal computer revolution, you probably can remember times when you began a new semester full of anticipation about the exciting new concepts in mathematics or science or engineering you were about to master. You may also recall that some of those times that sense of anticipation was quickly overwhelmed by disappointment and frustration as your delight in learning new concepts gave way to the drudgery of complicated and tedious calculations needed to apply these beautiful ideas to the all-too-messy real world.

If you had a kindly professor, like some of ours, you were probably told that faulty arithmetic would be partially excused (you would be given partial credit for your answer) since it was your understanding of concepts that counted, not the numerical value of the answer. Still, the ease with which a wrong minus sign, unit conversion, or other calculation flaw could creep into our work and rob us of the full sense of delight in our achievement made many of us complain about the dog work we had to do. What we would have given for an assistant who would do the computational drudgery and free us to gain enlightenment with less pain.

Even if your school was one of those fortunate enough to have a "comp center," you soon found that the early behemoths that were state of the art were anything but user-friendly. To do any useful mathematical work, you probably had to hike over to the computer center and (assuming you had mastered the complex computer programming languages then available) create some computer code and keypunch your program and data into computer cards. When you were done, being careful not to drop the cards and mix up the order, you gave them to the computer operator and returned in a few hours (if the computer was running properly) or the next day, to get a printout of your run. All too often, you found that there was a flaw in your program and you failed to get your calculation completed. So you went back to the keypunch machine and prepared to wait for another run.

Today, Mathcad can do the calculations and allow you to experience the joy of discovery that comes from learning new concepts of mathematics. This book aims to help you achieve the mastery that will allow you to use Mathcad with ease.

Howard Keller
John G. Crandall

Introduction

Plan of the Book

The book is intended to be an enjoyable tour through some of the most useful Mathcad features to enable you to quickly increase your skills at using Mathcad in your own work. After an explanation of what you need to know to use a certain feature, some examples of problems will lead you through how this particular feature will aid you in your own work. The initial examples will be quite simple so as to introduce a concept and show how Mathcad helps you deal with it. Then, the sample problems within each section become increasingly difficult as you progress through that section. By the time you reach the more challenging problems, you will realize that it is easy to do difficult work when you have the power of Mathcad behind you.

The authors have designed this book so that you may learn by doing. As you progress through the chapters, you will begin to see how some of the more advanced and powerful Mathcad features can be used for your own work. As you gain experience in working sample problems, you will see that even the power features are easy to use once you have a little practice. We hope you will try some of your own problems as you work through the exercises.

Keep in mind that this book is only one of the resources for learning to use Mathcad. The *Mathcad User's Guide* (MathSoft, 1997) and the on-line Help features (including the Tutorial, Help, Treasury, QuickSheets, Electronic Books, and other reference material located in the Resource Center) give a wealth of additional information and examples that can supplement what you learn from this book. This book is intended to help you get started, to give you a feel of how Mathcad works, and to show examples of how it can be used.

Finally, although this book is designed for use with all versions of Mathcad, you will find that some of the newest features can greatly enhance your performance and your pleasure in working with mathematics. So, if you have one of the older versions of Mathcad, this might be a good time for you to consider upgrading to version 7.

What Is New in Mathcad 7?

Mathcad has always featured powerful computing capabilities. Mathcad 7 both enhances the computational power and adds more ability to collaborate with colleagues and communicate about your work.

Some of the new and enhanced things Mathcad 7 can help you do are

- Use OLE 2 to drag and drop data from Mathcad to (or from) Word, Excel, or other OLE 2–compliant programs.
- Embed links to Excel or MATLAB™ to pass data to or from your Mathcad worksheet.
- Use the new MathConnex feature to create systems using data and components from various applications, including spreadsheets, Mathcad, and other calculation software. By using block diagrams, you can represent various components and data sets and link them together into a single system. The resulting system will present your work in a clear visual display that is self-documenting and live and that updates automatically. It then allows you to present your results without showing all the details of the calculations.
- Get hundreds of standard formulas from the Resource Center which you can drag into your worksheet.
- Make use of over 300 QuickSheets to help you perform many common tasks.
- Guides to practical statistics solving and programming.
- Extensive document preparation capabilities.
- Built-in Web connections and LAN (local area network) integration.
- New Windows 95 and NT interface including context-sensitive menus which you can access by clicking your right mouse button.
- Easier equation entry and editing.
- Smart conversion of units that automatically tracks and converts them for you and gives you complete SI (International System) units, as well as mks (meter-kilogram-second), cgs (centimeter-gram-second), and customary units.
- New data input and output enables you to move data in and out of Mathcad more quickly.

- Data filters for Excel files, MATLAB files, ASCII (American Standard Code for Information Interchange) files, and others.
- Extensive document preparation tools, including document templates, style sheets and region formatting, and improved page setup control and print preview.
- MAPI (massaging application programming interface)-based e-mail support.
- Define hyperlinks locally or to the Web.
- Free access to the "Collaboratory," a worldwide Internet forum for Mathcad users.
- Use Axum to access more 2D and 3D graphs and plots.
- Use Microsoft Internet Explorer 4.0, which is included free of charge, to enable you to access Mathcad and other sites on the Internet.

How Have the Menus Changed from Mathcad 6?

If you're upgrading to Mathcad 7 from Mathcad 6, you'll notice that the menu commands have been reorganized to make them more logical and intuitive. The "View," "Insert," and "Format" menus are new, and the "Text" and "Graphics" menus have been removed. Some of the other changes include

- "File/Export Worksheet" has been removed; you may now save a Mathcad worksheet in RTF (rich-text-formatted) format using the "File/Save As" command.
- "File/Open URL" (Uniform Resource Locator) has been moved to the Collaboratory dialog box. Access the Collaboratory with "File/Collaboratory," and use the URL text box at the bottom of the dialog box.
- The "File/Get from Notes" and "File/Save to Notes" commands have been removed. This functionality is no longer available.
- "File/Save Configuration" and "File/Execute Configuration" have been removed. You can now save preferred worksheet settings by creating a template.
- "Edit/Redo" has been added to let you reapply an edit undone by "Edit/Undo."
- "Edit/Delete" replaces "Edit/Clear."
- "Edit/Regions/View Regions" is now a command on the View menu.
- "Edit/Regions/Select All," "Edit/Regions/Separate," "Edit/Regions/Align Regions," "Edit/Regions/Lock Regions," and "Edit/Headers/Footers" have been moved to the Format menu.
- "Edit/Insert Pagebreak" and "Edit/Link" have been moved to the Insert menu.
- "Edit/Include" has been renamed "Insert/Reference."
- "Edit/Insert (Delete) Blank Lines" has been removed. Insert or delete blank lines by clicking in the worksheet and pressing the "Enter," "Backspace," or "Delete" key or click with the right mouse button in a blank spot and choose insert lines or delete lines to insert or delete more than one line at a time.
- The Text menu has been deleted, and the available commands have been distributed among

other menus. Insert a text region into your worksheet as before using the double-quote key " or choose "Insert/Text Region." The "Text/Create Text Paragraph" command is no longer available; only text regions can now be created. "Text/Embed Math" has been moved to "Insert/Math Region." "Text/Change Font" and "Text/Change Paragraph Format" have been moved to the Format menu. "Text/Check Spelling" has been moved to the "Edit" menu.

- The "Math/Matrices," "Math/Choose Function," and "Math/Units/Insert Unit" commands have been moved to the Insert menu.

- The "Math/Units/Change System of Units," "Math/Units/Dimensional Format," "Math/Randomize," and "Math/Built-in Variables" commands have been reorganized under the "Math/Options" pick.

- "Math/Numerical Format," "Math/Font Tag," and "Math/Highlight Equation" have been moved to the Format menu.

- "Math/Change to Greek Variable" is no longer available. Change a character behind the insertion point in math regions to Greek by pressing "[Ctrl]G."

- The Graphics menu has been deleted, and the available commands have been distributed among other menus. Use the "Insert/Graph" pick to insert available built-in two- and three-dimensional plot operators, and use the "Insert/Picture" command to insert a picture operator. Format a selected plot or picture by using the "Format/Graph" or "Format/Picture" command, or by double-clicking on the plot or picture. For X-Y and polar plots, "Trace" and "Zoom" are available as commands on Format/Graph as well as buttons on the graphing palette.

- The available "Symbolics" menu commands are identical to those in Mathcad Plus 6.0 but have been renamed or reordered for clarity. The "Differentiate on Variable," "Integrate on Variable," "Solve for Variable," and "Substitute for Variable" commands have been grouped under the "Symbolics/Variable" command. "Symbolics/Derivation Format" has been renamed "Symbolic/Evaluation Style," and "Symbolics/Derive in Place" is now part of the Evaluation Style dialog box.

- Most worksheet management commands formerly on the "Window" menu have been moved to the View, Insert, and Format menus. "Window/Zoom," "Window/Refresh," and the palette and toolbar hiding and viewing commands have been moved to the View menu. The "Window/Animation" commands have been renamed "View/Animate" and "View/Playback." The "Window/Change Colors" command has been moved to the "Format/Color" command.

- The "Books" menu has been removed. You can open an Electronic Book by choosing "Open Book" from the "Help" menu. To access the book annotation commands, open a book, and click on the book window with the right mouse button to see a context menu containing the annotation commands.

- The Help menu offers picks for "Mathcad Help," the "Resource Center," and "Using Help." The Desktop Reference and Tutorial Electronic Books have been integrated into the Resource Center, the Sampler Electronic Book has been removed, and Technical Support Help has been removed.

Tips for Using Mathcad 7

Mathcad 7 provides a "Tip of the Day" when you start to work. There are over 40 tips in all; you can cycle through them by clicking on "Next tip," or you can look at the whole file by searching for "tips" using "Find" on your Windows "Start" menu. You should find a file called "mtips" containing Mathcad tips and another called "tips" with tips for MathConnex. Make copies and format these tips files with Word's AutoFormat or a similar program to make them more readable—but leave the original file unchanged. Here are 10 Mathcad tips that the authors think are especially useful:

Tip 1.   For context-sensitive help on any function, click on the function name and press "F1."

Tip 2.   In Mathcad 7 you don't need to specify a range to plot a function. Just type the function using any unassigned variable, click on the *X-Y* plot button in the graphing palette, then click outside the plot. Mathcad uses -10 to 10 as the default range for the variable.

Tip 3.   Press "Space" to enlarge the selection in an equation.

Tip 4.   To move down from an exponent, press "Space" or the right arrow.

Tip 5.   To move out of the denominator of a fraction, press "Space."

Tip 6.   For help on an error message, press "F1" while the message is displayed.

Tip 7.   To shift the insertion bar from one side of a selected expression to the other, press "Insert."

Tip 8.   When editing math expressions, don't forget to check the status messages at the bottom of the screen for helpful suggestions.

Tip 9.   Mathcad's default unit for angle measure is radians. You don't have to type this unit. If you type "sin (3)," Mathcad assumes that you're using radians. To find the sine of 3 degrees (3°), type "sin (3deg)=." To turn an answer in radians into degrees, type "deg" in the placeholder to the right of the answer.

Tip 10.   To change the units in an answer, click once on the answer, click on the placeholder that appears to the right of the answer, and then type the name of the unit you want to use. When you click away, Mathcad will convert the answer to your chosen units.

Acknowledgments

The authors would like to express their thanks to the following people for assisting in the preparation of this book.

Glen Chatfield, President of Optimum Power Technology, for permission to use information in several exhibits.

Edward Adams, Frank Purcell, Mona Zeftel, and the others at MathSoft Inc. who were involved with the development of the concept for the book and the critiquing of the manuscript.

John Wyzalek, Steve Chapman, and Bernie Onken of McGraw-Hill for their invaluable assistance at all stages of publication.

Richard Keller and Shaun Keller for their contributions to the book.

Using the Collaboratory

MathSoft maintains an interactive World Wide Web service where Mathcad users can contact one another, post messages and read messages posted by others to exchange ideas about Mathcad, contribute Mathcad files, and download files posted by others.

The Collaboratory is located at *http://webserve.mathsoft.com/mathcad/collab/* and can be accessed by selecting Collaboratory from the File menu.

Using MathConnex

Mathcad Professional comes with a MathSoft Inc. application called MathConnex that allows you to build computational systems by linking Mathcad worksheets with other computational components such as Excel and Axum. In MathConnex, you draw lines that represent connection wires and tell MathConnex which outputs from components are to be used as inputs to other components. In this way you can chain together a number of components to produce very powerful and versatile computational procedures. A complete explanation of how to use MathConnex is beyond the scope of this book. You should read the *MathConnex Getting Started Guide* (MathSoft Inc., 1997) for a detailed explanation. You should also try some of the MathConnex sample problems that were probably installed on your computer along with the program. Look for a directory called "samples." The MathConnex sample problems are files with the .mxp extension. We suggest you start with Vdpol.mxp, which computes a solution to Van der Pol's Equation. To run MathConnex, click on the MathConnex icon from the Mathcad toolbar or exit Mathcad and run MathConnex as you would any application.

The Basics

What Is Mathcad?

Mathcad is the leading program for doing mathematics on a desktop computer. It is used by well over a half-million people, including scientists, mathematicians, college professors, students, engineers, financial analysts, MBAs, and many others whose work with numbers requires a powerful tool for mathematical and technical computations.

The first version of Mathcad (version 1.0) was released in 1986 to provide, for people who worked with mathematics, a product that could bridge the gap between the old-fashioned pencil and scratch pad and the newer electronic spreadsheet. Paper and pencil were versatile tools that allowed the user to display equations without artificial constraints imposed by hardware or software, but they required the user to supply a lot of mental input that was often drudgery of the least creative sort. Once the answer was calculated, changes to the problem required more drudgery. Spreadsheets avoided that problem by having the computer do the calculations, and the recalculations, once the relationships among the numbers had been established by the user. Recalculations were "live." No matter how complex the relationships among the numbers, whenever a number was changed, all the affected results were instantly recomputed. But spreadsheets hid the formulas behind the cells that displayed the data and were designed more for "profit-and-loss statements" and other accounting work than for equations and mathematical operations such as differentiation or integration.

The appearance of Mathcad (including the later versions) provided an elegant answer to this need by allowing users to input mathematical symbols and equations just as they would appear on a piece of paper, on a blackboard, or in a book. This allows the user to work with mathematics on a computer the way people think—which is not cell by cell. Furthermore, when the user changes one or more of the parameters in the equation, the "live calculation" aspect results in immediate recalculation of even the most complex results. For graphical presentations of the results, Mathcad immediately redraws the graph to show the new results. Experimenters can set up a problem on their computer, and quickly evaluate and graph the results as the values of the numbers in the equations are changed. Thus Mathcad provides an electronic version of the scratch pad—one which greatly enhances your ability to do your work, by allowing you to do the creative part of your job—while Mathcad takes care of the noncreative drudgery.

Mathcad has been acclaimed by numerous publications as a valuable tool for anyone who uses equations, or is trying to learn to use them. The various Mathcad versions have often been included in lists of the best available software, or selected as the outstanding programs of their type by computer magazines.

Although Mathcad has been upgraded with powerful new features several times, and has been made available for both Macintosh and UNIX, the original capabilities of the program have remained available to users of all versions. Because of this continuity, this book can be used by users of all versions of Mathcad. The 1994 introduction of version 5.0 was accompanied by a bifurcation of the product into two editions:

- The *Standard edition,* which included the basic capabilities required by someone learning mathematics or working on various types of engineering, financial, or scientific problems.

- The *Plus edition,* which included many enhanced features for solving advanced mathematical or scientific problems and for writing programs within Mathcad. In version 7, the Plus Edition became the *Professional edition.*

What You Need to Know to Use This Book

No prior knowledge of Mathcad is required in order to use this book. The book starts with very basic information and guides the reader through to more advanced concepts. Although some of the examples are based on mathematical or engineering or scientific concepts, it is not essential to know about these disciplines to use the book. The book does not attempt to teach any of these fields. Examples are drawn from various disciplines to provide more concrete illustrations and to suggest ideas about how readers might employ Mathcad in their own work.

The book does not assume that the reader has anything more than rudimentary computer knowledge. The book assumes only that you can turn the computer on and off, and get to Windows (or the operating system used by your computer). Most users should also know the basics of using the mouse, since it will be used for selecting items from menus and palettes.

How to Use This Book

Although Mathcad is relatively easy to use, some features may seem more intuitive after you have been guided through some examples. You are encouraged to try the exercises in the book (they are also included on the CD-ROM) and change the values to see how Mathcad recalculates the results. Also, think of how these examples relate to work you do. As you develop mastery of Mathcad, please experiment and try some of your own problems.

Scope of This Book

This book is intended as a guide to using the features that are most likely to help you get results in your work. It is not intended to be an encyclopedic reference to every one of the vast number of Mathcad features; nor is it intended to replace the *Mathcad User's Guide*. It is also not a substitute for scientific or mathematics textbooks.

Although the authors have used the latest Mathcad version (version 7) in producing this book, it is not their intention to explain every one of the new features. The *Mathcad User's Guide* is the best reference source for a comprehensive listing of both old and new features. We emphasize features that we believe will help you get up to speed most quickly.

Installing Mathcad

The installation procedure will depend on your particular computer setup and version of Mathcad. This procedure is described on the Mathcad 7 CD insert. For the Windows 95 (or Windows NT 4.0) operating systems, being used by the authors, you insert the first Mathcad floppy disk or CD-ROM into the appropriate drive and choose "settings" from the startup menu. Then, select "control panel" from the extension of the startup menu. Finally, click on the "add/remove software" icon in the "control panel" window, and follow the instructions given.

Starting Mathcad

From the Windows 95 startup menu, select "Mathcad 7.0" and then select "Mathcad 7 Professional" from the Extension menu—assuming that you are using Windows 95 and version 7 Professional. For other combinations, refer to the CD insert.

Mathcad will now start with a new document screen and a red cursor in the form of a cross, indicating where your data entry will begin. To move the cursor, in order to start your data entry elsewhere, simply move your mouse cursor and click at the desired spot.

How the Examples Look

Although many of the basic features are the same in various versions, the appearance of the screen will vary depending on which version and operating system you use. The authors are using personal computers with Pentium chips and Microsoft's Windows 95 or Windows NT 4.0 as the operating systems and Mathcad 7 Professional for the examples. Although we may occasionally describe how the example would be done differently on a different version of Mathcad, we have not tried to describe every possible permutation of operating system and Mathcad version. In most cases, you should have no difficulty following our examples on your computer.

Getting Acquainted with the Mathcad Data Entry Screen

Begin by looking at Exhibit 1-1, which is a screen print of the Mathcad screen as it appears for version 7 Professional running in the Microsoft Windows 95 operating system.

Exhibit 1-2 is another screen print for comparison. This is the previous version, 6.0 Plus. Note that the basic features are the same, but the arrangement of icons, tools, and palettes is somewhat different.

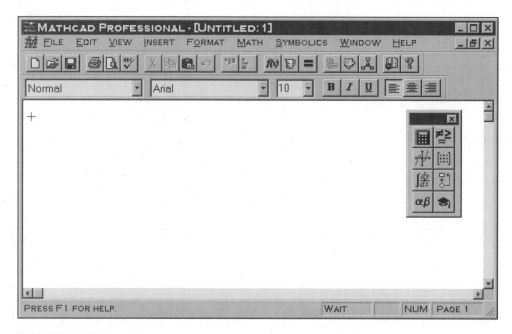

Exhibit 1-1 Main screen.

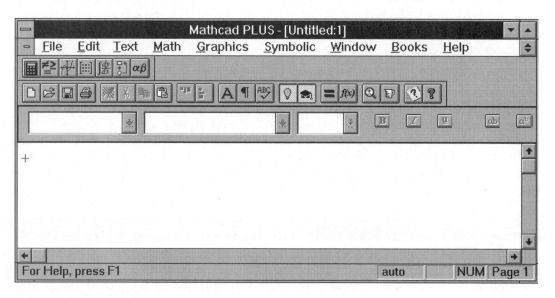

Exhibit 1-2 Main screen (version 6).

At the top of the Mathcad 7 Professional screen, just below the Title bar, you will see the Windows Menu bar with such familiar entries as "File," "Edit," "View," "Window," and "Help." If you have some previous experience with Windows, you will recognize these as the menus you use for such basic "housekeeping" duties as getting, saving, and printing files; editing documents; or controlling the appearance of the windows on your screen. You will also see some menus more specific to Mathcad:

Insert

Format

Math

Symbolics

We will learn more about these Mathcad menus later, but first let us take a brief look at each of the menus. You may want to try using your mouse to pull down each of these menus for a quick inspection.

The "File" menu, shown in Exhibit 1-3, allows you to do basic file management tasks such as creating and saving a new Mathcad file or opening, closing, or saving one you used previously. Other actions you can select from the File menu include setting up and gaining access to the Internet (from within Mathcad), or accessing Mathcad's Collaboratory or sending Mathcad worksheets by e-mail (provided you are connected to an e-mail system that is compatible with Microsoft Mail). You can also set up your worksheet pages (portrait or landscape mode, etc.) and print your worksheet, or preview how the page will look when printed. At the bottom of the File menu, Mathcad displays a list of some files you have worked on recently. You can open one of these files by using your mouse to click on the filename in this list.

The "Edit" menu, shown in Exhibit 1-4, gives some basic editing capabilities such as cutting data from a file and pasting into another place in that file or another file, deleting selected parts of the worksheet, or selecting the entire worksheet. An "Undo" is also available at the top of the Edit menu, in case you make a mistake or change your mind. Undo reverses the last action—although some actions cannot be reversed (in which case the Undo will be grayed out and thus be unavailable). "Redo" reinstates an action that you have undone. "Find," "Replace," and "Go to Page" are available to help you move to the part of your Mathcad document where you want to make changes. "Check Spelling" starts Mathcad's technical spell checker, which includes many mathematical and technical terms that would not be recognized by ordinary spell checkers. "Links" and "Object" are used in working with data from other Mathcad documents or (in the case of "object") other applications that you want to use, or make reference to, in this worksheet.

The "View" menu, shown in Exhibit 1-5, allows you to control the look of your Mathcad screen. The "Toolbar," "Format Bar," and "Math Palette" toggle these parts of the screen on and off each time you select them, enabling you to have a cleaner, uncluttered screen or to make these tools available as needed. "Regions" shows the math and text regions of your document; "Zoom" controls the magnification of the image you see on your computer screen, allowing you to enlarge the image to see more detail or to make it smaller and see more of the worksheet at one time. "Refresh" is used to redraw the screen image after extensive editing. "Animate" and "Playback" are used to create animated graphs of your results.

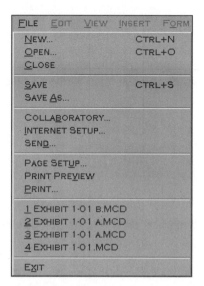

Exhibit 1-3 File menu.

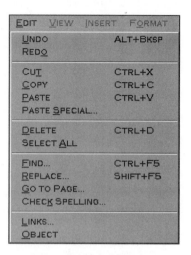

Exhibit 1-4 Edit menu.

The "Insert" menu, shown in Exhibit 1-6, allows you to insert graphs, matrices, mathematical functions, or units (pounds, kilometers, etc.) into your work. You can also create "Math Regions" or "Text Regions" in your worksheet or insert a "Page Break" (to start a new page at the insertion point) from this menu, or you can enhance the appearance of your work by adding a picture. This menu has a submenu (shown in Exhibit 1-7) that enables you to

Exhibit 1-5 View menu.

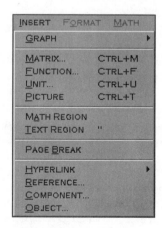

Exhibit 1-6 Insert menu.

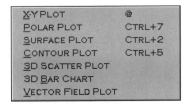

Exhibit 1-7 Insert—Graph submenu.

Exhibit 1-8 Insert hyperlink submenu.

insert graphs, and two submenus (see Exhibits 1-8 and 1-9) for inserting hyperlinks into your Mathcad documents.

The "Format" menu, shown in Exhibit 1-10, makes it easy to control the way your Mathcad work will look on the screen or when printed. Formatting has been significantly enhanced in version 7. You can format the display of your numbers, equations, text, paragraphs, styles, and graphs. You can specify colors to enhance the appearance and clarity of your documents. You can also control your headers and footers and clean up the alignment of text and math regions using this menu. "Properties" provides the capability to highlight with color, or place a border around specified equations or text and lock parts of your work. The submenus are shown in Exhibits 1-11 to 1-14.

The "Math" menu, shown in Exhibit 1-15, allows you to control how Mathcad does calculations. The default setting is "Automatic Calculation" mode, which produces Mathcad's famous live calculations. You get results immediately without having to tell Mathcad to calculate or

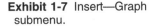

Exhibit 1-9 Insert hyperlink.

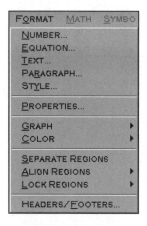

Exhibit 1-10 Format menu.

Exhibit 1-11 Format— Graph submenu.

worrying about how the calculations are done. "Calculate" and "Calculate Worksheet" allow you to have Mathcad turn off "live" calculations and wait until you are ready, before doing the calculations. At times you may want the symbolic processor and the numeric processor to work together for better results. "Optimization" allows you to tell Mathcad to have them work together in order to produce the optimal mathematical technique for doing your calculation. "Options" gives you control over the units system to be used, the dimensions, and some internal tolerances and printing parameters.

The "Symbolics" menu, shown in Exhibit 1-16, allows you to do calculations that result in answers that are mathematical expressions as opposed to specific numbers, or that give exact answers instead of numerical approximations. The submenu for symbolic calculations on a variable is presented in Exhibit 1-17; for symbolic matrix calculations, in Exhibit 1-18; and for transforms, in Exhibit 1-19.

The "Window" menu, shown in Exhibit 1-20, gives you the ability to arrange the windows and icons on your screen. Perhaps most important, the Window menu lists the open Mathcad

Exhibit 1-12 Format— Color submenu.

Exhibit 1-13 Format— Align Regions submenu.

Exhibit 1-14 Format— Lock Regions submenu.

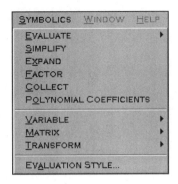

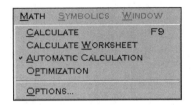

Exhibit 1-15 Math menu.

Exhibit 1-16 Symbolics menu.

documents (at the bottom of the menu), enabling you to easily move from one document to another by selecting from the list.

In addition to giving you access to Mathcad's on-line help, the "Help" menu, shown in Exhibit 1-21, gives you lists of all the keyboard commands, gives you access to QuickSheets (a helpful feature explained later in this book) and other useful problem-solving aids in the new Resource Center. Go to your Electronic Books, optional on-line guides to specific types of problems you can purchase from MathSoft, through the "Open Book" command. Tips about using Mathcad are accessible through "Tip of the Day." To see what version of Mathcad you are using, select "About Mathcad," which is located at the bottom of the "Help" menu.

Either below the Menu bar, or freestanding nearby, you will see a palette with a group of seven icons with pictures related to mathematics on their faces. These icons are used to call up (by means of a mouse click) palettes that allow you to click on mathematical symbols or operators or Greek letters and thereby have easy access to them for entry into your work—since some of these are not found on standard computer keyboards. You can drag these palettes around the computer screen with your mouse in order to be closer to your work or to the top of the document window where they will become a palette toolbar near the top of the screen. Keystroke combinations are also available for many of these symbols for users who prefer

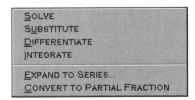

Exhibit 1-17 Symbolics— variable submenu.

Exhibit 1-18 Symbolics— matrix submenu.

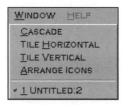

Exhibit 1-19 Symbolics—
transform submenu.

Exhibit 1-20 Window
menu.

working from the keyboard (instead of using the mouse). A list of these keystrokes is included in the *Mathcad User's Guide* or from the Help menu.

The "Arithmetic" button and palette are shown in Exhibit 1-22. The Arithmetic palette has the digits 0 through 9. Although these are on standard keyboards, Mathcad gives you another way to access them since these are frequently needed for your equations, and clicking your mouse on the palette may sometimes be preferable to moving your hands between the keyboard

Arithmetic Button

Arithmetic Palette

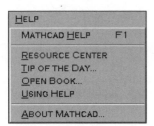

Exhibit 1-21 Help menu.

Exhibit 1-22 Arithmetic
button and palette.

and the mouse. Also on this palette are some common mathematical constants, parentheses, some basic trigonometric functions, the basic arithmetic operations, some powers and roots operators, and a few other frequently used symbols and operations. By having all these on a palette, Mathcad makes it easy for you to enter any of these into your work with the click of a mouse.

The "Evaluation and Boolean" button and palette are shown in Exhibit 1-23. This palette includes one of the frequently used operators. It is prominently located in the top row of the palette and is used to tell Mathcad to do a symbolic calculation. It is the right-pointing arrow. This palette also includes symbols used for globally assigning values and for creating inequalities, boolean expressions, and several types of operators. If you don't know what all these are used for, do not worry. We will learn more about them later on.

The "Graph" button and palette are shown in Exhibit 1-24. This palette allows you to create graphs of your data with the click of a mouse. This is something we will be doing a lot of later. With Mathcad's graphical capabilities at your service, painstakingly drawing plots of your numbers on graph paper is something you will never have to do again. Furthermore, you will be able to communicate your answers in a clear and lively manner to your readers. You can use this palette to create an instant graph of a function by typing the function definition (for example, x^2) and clicking on the graph at the top left of the Graph palette.

The "Vectors and Matrices" button and palette are shown in Exhibit 1-25. This palette gives you operations for creating and using vectors and matrices. This is another palette that will be used often, further on in the book.

The "Calculus" button and palette are shown in Exhibit 1-26. This palette gives operators

Graph Button

Evaluation and Boolean Button

Evaluation and Boolean Palette

Graph Palette

Exhibit 1-23 Evaluation and Boolean button and palette.

Exhibit 1-24 Graph button and palette.

Calculus Button

Vectors and Matrices Button

Calculus Palette

Vectors and Matrices Palette

Exhibit 1-25 Vectors and
Matrices button
and palette.

Exhibit 1-26 Calculus
button and
palette.

you will use for calculus problems such as limits, integrals, derivatives, and summations. This is another important set of operators that you will learn how to use.

The "Programming" button and palette are shown in Exhibit 1-27. This palette, available only in the Professional edition, is used for programming. It was a new feature in the version 6.0 Plus and has been enhanced in the Professional edition of version 7. We will not be concerned about this until later on in the book.

The "Greek Symbol" button and palette are shown in Exhibit 1-28. This palette helps you create equations which look just as they do in textbooks. It allows you to easily enter letters of the Greek alphabet (lowercase and uppercase) into your mathematical expressions without remembering combinations of keystrokes. So your equations can have real alphas, betas, and other Greek letters that allow you to write mathematics as traditionally done before computers came into existence.

The "Symbolic Keyword" button and palette are shown in Exhibit 1-29. This palette makes available a number of operations that are used in symbolic mathematics and will be discussed in more detail later in this book.

If you click on the "Modifiers" keyword on the "Symbolic Keyword" palette, you get the secondary palette shown in Exhibit 1-30. These modifiers will be discussed later.

Greek Symbol Button

Greek Symbol Palette

Programming Button

Programming Palette

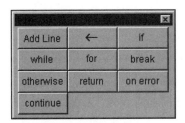

Exhibit 1-27 Programming
button and palette.

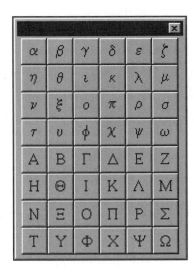

Exhibit 1-28 Greek Symbol
button and palette.

Below the menus on the Mathcad screen are two toolbars that can be turned off and on from the View menu. The toolbar contains such Windows-standard tools as those that open, save, print, or edit a file as well as Mathcad-specific tools for entering mathematical functions or units or use the Resource Center or MathConnex. The tools shown in Exhibit 1-31 are used to align regions, text, or mathematics to improve the appearance of your documents. Exhibit 1-32 shows a tool used for inserting units, such as miles, pounds, or volts, into your work. Exhibit 1-33 inserts mathematical functions, from Mathcad's vast library of them, into your work. Exhibit 1-34 shows a tool that tells Mathcad you have finished entering your mathematical expression and are ready for Mathcad's numerical processor to go to work and calculate the answer. You can also type an equal sign "=" to accomplish the same thing. A different symbol functions as a "symbolic equal sign." The "Insert Hyperlink" tool (Exhibit 1-35)

Symbolic Keyword Button

Symbolic Keyword Modifiers Button

Symbolic Keyword Palette

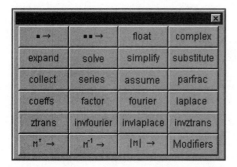

Symbolic Keyword Modifiers Palette

Exhibit 1-29 Symbolic Keyword button and palette.

Exhibit 1-30 Symbolic Keyword Modifiers button and palette.

allows you to put links in your Mathcad work. The "Component Wizard" (seen in Exhibit 1-36) provides an easy way to insert work from MATLAB, Excel, Axum and other components into Mathcad.

The tool shown in Exhibit 1-37 starts MathConnex. The tool shown in Exhibit 1-38 can be used to call up Mathcad "Help."

The "Format Bar" includes tools that control the font, font size, and left or right adjustment

Align Across

Align Down   Insert Unit   Insert Function

Exhibit 1-31   **Exhibit 1-32**   **Exhibit 1-33**

Calculate

Exhibit 1-34

Insert Hyperlink

Exhibit 1-35

Component Wizard

Exhibit 1-36 Tool used
to insert other
components into
Mathcad.

---

as well as other aspects of your worksheet's text appearance. If you are familiar with a previous version of Mathcad, you will notice that there are major enhancements to Mathcad's formatting capabilities in version 7. You can now use different formatting in parts of your text within one text region—for example, making part of your text a different size or different color using this text bar and the Format menu. This will enable you to highlight parts of your text and prepare better-looking documents that can more effectively communicate your work.

Now that you have acquired some familiarity with the tools and capabilities available to you when you start up Mathcad, you are ready to learn to use some of the basics.

Basic Features and Capabilities

Mathcad is designed primarily for mathematical solutions to your problems, but Mathcad's developers knew that it is also important that you be able to explain your work in words. So Mathcad incorporates text-processing capabilities that allow you to document and annotate your mathematical work and communicate your ideas to your coworkers and colleagues at other organizations. You can even write your technical and mathematical papers in Mathcad and incorporate "live" equations.

Getting Started

As you begin your work, Mathcad has opened a blank document on your Mathcad screen. A new document looks like a white sheet of paper. If you see a gray (or other nonwhite color) screen and no red cursor, you do not yet have an open document. You will need to open a new document before

---

Run MathConnex

Exhibit 1-37 Tool used
to start MathConnex.

Help

Exhibit 1-38 Tool used
to call up Mathcad
Help.

you begin to work. Go to the File menu and select "New" or use the "New Document" icon at the far-left end of the toolbar. Once you have started a document, you can begin using Mathcad. As befits a mathematics program, the screen for a new document is initially set for you to enter numbers, mathematical symbols, and variable names—including letters of the Greek alphabet. As we saw earlier, Mathcad has provided palettes and keystroke combinations for entering letters and symbols that are not standard keys on a computer keyboard.

To enter text, you must "tell" Mathcad that what you are about to key in is to be understood as text so that the program will not try to use it to calculate. This is easily done in Mathcad by selecting "Text Region" from the Insert menu. When you do this, Mathcad produces a small region on the screen that is intended for the input of text. As you type, Mathcad accommodates you by continually expanding the area designated for text until you have finished typing. In version 7, Mathcad has learned to recognize text. When you begin typing a text entry, Mathcad starts a text region without the necessity of you telling it to do so.

Now that you have entered text, to get back to mathematics, simply move your mouse outside the text that you have just typed and click the left mouse button. Once again you will see the familiar red cursor that specifies where your next mathematical entry will appear on the screen.

The distinction between text and mathematics is important to understand because Mathcad treats them very differently. Mathematical entries cause Mathcad to do calculations, or give you an error message if it cannot calculate the entry as you have entered it. Text, on the other hand, is displayed by Mathcad, but otherwise not used. Mathcad does not try to evaluate your text, even if you type numbers or mathematical symbols in text. On the other hand, Mathcad's developers know that it is sometimes useful to be able to insert live equations into a text region or paragraph, so they have given you a way to do so. If you copy a live equation from another part of your document and insert it into a text area using the "Math Region" command from the Insert menu, the live equation will continue to be live even though it is no longer in a mathematical area of the screen. You could, for example, explain a formula for computing the payment amount for a loan and enter the formula into the explanation. The reader could now enter some data into the formula and immediately get the correct answer as calculated by Mathcad. To copy or insert equations into text:

- Click in text
- Choose Math Region from Insert Menu
- Type or paste

How to enter and change data

You have seen how to enter text, but now you need to enter numbers and mathematical symbols. Entering numbers is easy. Just make sure your cursor is in the place where you want the number to appear on the screen, then use the number keys at the top of your keyboard or on your number pad (located on the right end of the keyboard on most keyboards) and the period "." key to type the decimal point (for noninteger numbers).

How to select and edit

If you are already familiar with the use of your computer for other programs, you will find that the techniques for selecting and editing mathematical data in Mathcad are a little different than selecting and editing in most Windows or Macintosh programs. We will see how to select and edit as we proceed through this chapter.

Remember, if you need help at any time, press the F1 key (found in the top row of keys on many keyboards). In general, this will give you access to the full on-line help, and you can select the topic of interest. If, however, you have your cursor in a mathematical expression or text region at the time you press the F1 key, you will activate "context-specific help," which means that Mathcad will give you selected help information that it "thinks" will be useful for the work you are doing now. For example, if you type in "e^2" and then press the F1 key, Mathcad will give you a help topic that discusses "exponentiation" and gives you links to related topics. You can also get context-specific help about menu topics by opening a menu and putting your cursor on the relevant selection from that menu—Mathcad will give you a message at the bottom of the screen, describing briefly what that command will do. You have also seen how using your mouse to move the cursor to a palette, or a button on a palette, causes Mathcad to display a brief description of that palette or button.

If you have used Mathcad previously, you will immediately discover that entering and editing mathematical expressions has been improved in version 7. The new method is simpler and more intuitive. When you start to enter a number or a mathematical expression, Mathcad creates a special two-part cursor, with a horizontal line under the entry and a vertical line to the right of the entry. The whole entry is also surrounded by a rectangle while you work on it. To get rid of the cursor, move the mouse outside the expression and click. The basic idea about the use of this cursor is that the two lines envelop the part of the equation you are working on at the time (exponent, numerator, denominator, both nominator and denominator, etc.). To work with the whole equation, you need to enlarge the horizontal and vertical lines to enclose the whole equation, but you would want to move the cursor lines to the exponent in order to work with it exclusively. You can change the size and position of the cursor by using the space bar or the arrow keys (also called *cursor keys*) on your keyboard. As in the mathematics you learned in school, you can enter parentheses to control the order of execution of the mathematical. If you want to change something at the left of the expression, use your left-pointing arrow (or cursor) key.

When you place your mouse cursor in a mathematical expression you had previously created in Mathcad and click the left mouse button, you will see the same vertical and horizontal lines making up the cursor as you saw when you entered mathematical information from scratch. The right mouse button will bring up a small menu of commands such as "cut" or "copy" that you can use in editing the expression. If you use the backspace key, the cursor will move to the left and erase a digit of a number you have entered or a mathematical operator. To select part of a mathematical expression, click on your spacebar. You will now see editing lines in the mathematical expression. Clicking additional times will expand the scope of the editing lines. In Mathcad, you use this selection procedure not only to cut and paste data as you would in other types of programs but also to tell Mathcad what part of the expression you want to edit: the exponent, the numerator or denominator, and so on. You are now ready to begin using Mathcad to do some work.

When you have typed a mathematical expression and want to evaluate it, you just left click somewhere in the expression, and type an equal, "=," sign. Almost magically, Mathcad will calculate and display the answer.

If you want to drag an expression you have entered (in its rectangular container) to a new location on the screen, move your mouse around the rectangle until the cursor shaped like a hand appears, then drag the whole rectangle to the new location.

If you want to delete an expression you have entered, drag the arrow cursor through the rectangle until black highlighting appears, then press the "Delete" key. Alternatively, expand the cursor to include the whole expression, within the rectangle, then move the vertical part of the cursor to the front of the expression. If you now press the Delete key, the black highlighting will appear, and pressing the Delete key again, will delete the expression. You can also highlight part of the expression and delete that part.

Basic calculations

Despite Mathcad's considerable mathematical power, you should not feel intimidated. You will find that as soon as you have opened the package and installed the software, you are ready to use some of its basic features. Mastery of some of the more advanced capabilities will require additional time and experimentation. This book is intended to help you with that process as you proceed to the later chapters, but right now we will see how easy it is to get started with the basics.

If you are new to computers, you should be aware that Mathcad is very patient. Feel free to take your time and experiment until you understand what Mathcad expects you to do. Also, realize that Mathcad will not criticize you for making a mistake or for not understanding how to do something. You are here to learn, and Mathcad is an ideal partner in learning, because it allows you to experiment and make mistakes. Although Mathcad may sometimes give you an error message, if you do something that it does not understand, it will not criticize you for your mistake, and you should not get discouraged or frustrated. Just keep trying for as long as you wish until you begin to understand how Mathcad "thinks." In addition, you should realize that any mistakes you make in using Mathcad will not damage your computer or get you in any other serious trouble. Just about the worst thing that could happen as a result of a mistake is for your computer to "lock up." If that were to happen, as with any other program, you should restart your computer and resume your work where you left off. Remember to save your work occasionally to avoid a major data loss just in case you should have such a problem.

Begin by trying some simple calculations

Mathcad's screen can function as a very powerful calculator. For example, let's assume that you want to calculate a problem in basic arithmetic, such as adding two numbers.

For this first example we will use an extremely simple problem: how to add 1 plus 2. Type in the first number, "1," followed by a plus sign, "+," then the second number, "2." After you have typed the expression "1 + 2," type an equal sign, "=." Mathcad instantly calculates the answer, "3." (See Exhibit 1-39.)

Now we will do some other very simple examples, to further illustrate how Mathcad can

$$3.754 + 2.831 + .011 = 6.596$$

And

$$1 + 2 = 3$$
$$12345698.99 + 33235 + 66 = 1.238 \cdot 10^{7}$$

Exhibit 1-39 Addition.

Exhibit 1-40 Addition.

function as a calculator. Most of your work will be much more advanced than these examples, but it is useful to learn how to use Mathcad to perform even very simple calculations. Later, when we look at how to use vectors, you will see some other ways to add numbers. In the meantime, two more examples of addition are given in Exhibit 1-40.

You have probably already noticed that Mathcad displayed the answer to the third addition example in scientific notation (i.e., a number with a decimal point, followed by 10 raised to a power). Further on in this book, there will be more information about how Mathcad displays results and how you can set the format of the answer.

By now, you may have already realized how to do subtraction. It is similar to addition, but you just substitute a minus sign, "−," instead of the plus sign, "+," when you type in the expression. This is followed by the equal sign, "=," to get the result.

Multiplication is similar to addition and subtraction, as you might expect. You type in the numbers separated by a multiplication operator, which in Mathcad is the asterisk, "*." You will find the asterisk just above the number "8" on your keyboard—so you must use the shift key, together with the "8." Notice that although you typed an asterisk, the asterisk becomes a dot (signifying multiplication) in a Mathcad formula, in accordance with standard mathematical notation. Type an equal sign following the expression to get the result. Exhibit 1-41 illustrates some examples of subtraction and multiplication.

So far, our results are pretty much the same as what you would get if you were using a typewriter or word processor. But when you divide, you begin to see one of Mathcad's most important features. In Mathcad, equations look just as they would if they were written on a blackboard, on paper, or in a book.

$$2300 - 200 = 2.1 \cdot 10^{3}$$

$$11.549 - 6.3 - 2 = 3.249$$

$$\frac{5}{7} = 0.714$$

$$2 \cdot 3 = 6$$

$$1000000 \cdot 1000000 = 1 \cdot 10^{12}$$

$$\frac{5 \cdot 6}{2} = 15$$

Exhibit 1-41 Subtraction and multiplication.

Exhibit 1-42 Division.

To divide, type in the numerator, followed by the slash key, "/" ("/" represents "divide"). Then, follow this by an equal sign. Mathcad puts the numerator above the denominator, separated by a horizontal line—just as you would do if you wrote this fraction on paper. So far, we have used only numerical problems in these examples, but later we will use mathematical expressions. You will also learn about Mathcad's techniques for entering more complex mathematical symbols and expressions.

Are these results live?

You have already learned that one of Mathcad's important features is "live recalculation." The work you have just done is live. To illustrate what that means, go back to the problem we have just completed, and use the mouse to click somewhere in the problem. You will see a cursor where you just clicked. If you insert the cursor just after a number you want to change, you can now edit the problem by using the backspace key to erase all or part of the number. Then, type a different number in its place. Immediately, the answer is updated to reflect the new data. For example, click just to the right of the "6" in the last example and backspace to erase the "6." Now, type a "12" where the "6" used to be and watch as Mathcad recalculates the answer. The previous answer, "15," is replaced by the new answer, "30." You may not be particularly impressed with this now because this is such a simple problem, that you could easily calculate the answer in your head. However, later, you will discover in much more complex problems how truly impressive this capability is.

Some Slightly More Complex Calculations

Now we do a few more examples. These still use only basic mathematical operations, but the problems combine a few more steps to give you some experience in entering more complicated math. For these examples we need to see how to enter equations involving powers and roots. The key on your keyboard for this operation is the caret, "^." If you enter a fraction or decimal fraction, Mathcad will compute the specified root of the number. Otherwise, it will raise the number to the stipulated power.

The arithmetical palette gives you an easy way to enter such calculations into your work with a mouse click. Click on a symbol to calculate square roots, nth roots, x^{-1} (inverse), and x^y. Once you enter one of these operators into your work, you will see one (or more) little black registers called *placeholders*. These are Mathcad's way of telling you that it is looking for some input from you. Just click on the rectangles and enter your data. In most cases, it will be readily apparent what type of input is being requested. Once you have done a few more of these, we will look at some problems that demonstrate some basic calculations of the type someone might do in their real work. Now, please refer to Exhibits 1-43 through 1-53.

Calculations involving parentheses

Mathcad "knows" the rules of precedence that are used in mathematics (i.e., powers and multiplication or division are done before addition and subtraction, etc.) and uses them in its calculations. Mathcad also knows that you sometimes want to override the default rules by

$$\frac{1}{777} = 1.287 \cdot 10^{-3}$$

$$777^{-1} = 1.287 \cdot 10^{-3}$$

Exhibit 1-43

$$\frac{21.2^2 \cdot 8.95}{17.6 \cdot 61.7 \cdot 4.6} = 0.805$$

Exhibit 1-44

$$\sqrt{605000} = 777.817$$

Exhibit 1-45

$$0.043^2 = 1.849 \cdot 10^{-3}$$

Exhibit 1-46

using parentheses to group operations the way you want them to be performed. Remember to move the vertical cursor line right and left by using the right-pointing and left-pointing cursor keys on your keyboard. In Exhibits 1-54 and 1-55 we show a couple of examples of the use of parentheses in Mathcad. We are still using only very basic operations in this section. Your real work may be much more complicated, but these problems will show the basics of using parentheses in Mathcad.

Basic time calculation

Here, we have an example of something else you can do using the operations you have learned so far. A compact-disk recording of music has the timings for the music tracks shown in Exhibit 1-56. You want to get the total time for the recording by adding the times for the tracks, but you have to take into consideration that these are times (shown in minutes and seconds—not decimals) and convert to decimals before doing the calculation.

$$\sqrt[3]{.041} = 0.345$$

And

$$\sqrt[5]{.041} = 0.528$$

To check this

$$.528 \cdot .528 \cdot .528 \cdot .528 \cdot .528 = 0.041$$

Or

$$.528^5 = 0.041$$

$$\sqrt{0.00094} = 0.031$$

Exhibit 1-47

$$\sqrt[3]{0.32} = 0.684$$

Exhibit 1-48

Exhibit 1-49

$\sqrt[3]{.0075} = 0.196$

And

$\sqrt[7]{.0075} = 0.497$

And
If you want to compute non integer roots

$.0075^{\frac{1}{3.5678}} = 0.254$

To Check

$$\sqrt[3]{.05} = 0.368$$

$.254^{3.5678} = 0.008$

And

This answer is rounded to three decimal places
Later we will see how to set the displayed precision

$.05^{\frac{1}{3}} = 0.368$

Exhibit 1-50

Exhibit 1-51

Government budget

An article showing funds allocated for various types of services by the U.S. federal government (for fiscal year 1996) includes the amounts shown in Exhibit 1-57. Later, we will show how to use arrays (vectors and matrices) to handle groups of numbers such as these, but at this early stage we take a more straightforward approach using the "add" operation (as represented by the plus sign, "+."

Masses of the sun and planets
(in kilograms)

Exhibit 1-58 shows the masses of the earth and the sun in kilograms. We will use these metric units in the next section. Here, for those who do not want to use metric units, we explain how you would convert a planetary mass, stated in kilograms, to tons. You could use the old-fashioned technique of multiplying by a conversion factor to change from one set of units to another, but Mathcad has an easy-to-use built-in capability to work with units and convert units. Unit conversion is demonstrated in this exhibit.

Without parentheses

$$2.5 + 7 \cdot 5 + 10 = 47.5$$

With parentheses

$$\frac{30.6 \cdot 41.2 \cdot 5.41}{40.8^2 \cdot 7.3} = 0.561$$

Exhibit 1-52

$$\frac{100 \cdot 60.5^3}{48 \cdot 3 \cdot 10^4 \cdot 655} = 0.023$$

Exhibit 1-53

$$(2.5 + 7) \cdot 5 + 10 = 57.5$$

Exhibit 1-54 Parentheses in Mathcad.

Gravitation

In this section we see how the gravitational force between any two objects is calculated from the law of gravitation discovered by Newton. We calculate (Exhibit 1-59) the gravitational force between the earth and the sun using data from the previous section. You can easily extend this example by substituting the mass of another planet instead of the earth. (You also need to insert the correct distance between the two objects in the denominator of the equation.) The distance is squared in the calculation. You can also use the masses of two planets (instead of the sun and a planet), and thus you can calculate the gravitational pull between these planets. You must remember to insert the correct distance between the planets in the squared term in the denominator. Finally, remembering that Newton's gravitational formula has very general applicability, you can enter the masses and distance separating any two objects (cars, people, etc.) into the equation and calculate the gravitational force between the two objects. Of course, for ordinary objects, the masses will be much smaller than for bodies of the solar system, so the gravitational force will be correspondingly smaller.

Without parentheses

$$\frac{1}{2} + .3^2 = 0.59$$

With parentheses

$$\frac{1 + .3^2}{2} = 0.545$$

We typed parentheses to group the numerator, but they disappeared as soon as we typed /, since they were no longer needed.

Exhibit 1-55 Parentheses in Mathcad.

The time for the tracks of a compact disk are as follows:

	Minutes and seconds (shown on package)	Entered into Mathcad as
Track 1	8:11 minutes	$T1 := 8 + \dfrac{11}{60}$
Track 2	9:43 minutes	$T2 := 9 + \dfrac{43}{60}$
Track 3	3:00 minutes	$T3 := 3$
Track 4	6:49 minutes	$T4 := 6 + \dfrac{49}{60}$
Track 5	4:23 minutes	$T5 := 4 + \dfrac{23}{60}$
Track 6	30:20 minutes	$T6 := 30 + \dfrac{20}{60}$

Total = $T1 + T2 + T3 + T4 + T5 + T6 = 62.433$

Or a total of   62   minutes and

26   seconds (approx.)

There is an alternative, and more elegant, way of doing this problem that requires knowledge of techniques from a later chapter.   We include it here as a preview of what lies ahead.

$$M := \begin{bmatrix} 8 & 11 \\ 9 & 43 \\ 3 & 00 \\ 6 & 49 \\ 4 & 23 \\ 30 & 20 \end{bmatrix}$$

Then the total time in minutes is

$$\sum M^{\langle 0 \rangle} = 60 \qquad \sum M^{\langle 1 \rangle} = 146$$

Or   62 minutes + 26 seconds

Exhibit 1-56 Time calculation.

An article about a proposed budget plan for the U.S. Federal Government gave the following amounts in ($ billions)

Receipts

Individual Income Taxes	645.1
Social Sec. & Medicare Taxes	536.2
Corporate Income Tax	185.0
Excise Taxes	59.6
Miscellaneous	31.8
Custom Duties	20.8
Estate & Gift Taxes	17.1

Total Receipts $645.1 + 536.2 + 185.0 + 59.6 + 31.8 + 20.8 + 17.1 = 1.496 \cdot 10^3$

So the total of the budgeted receipts is $1,496 billions or $1.496 trillions

Outlays

Social Security	398.1
Treasury	368.1
Health and Human Svs.	354.3
Military	247.5
Agriculture	55.9
Office of Personnel Mgmt.	44.6
Veteran's Affairs	39.8
Transportation	38.1
Labor	35.2
Other	58.4

Total Outlays

$398.1 + 368.1 + 354.3 + 247.5 + 55.9 + 44.6 + 39.8 + 38.1 + 35.2 + 58.4 = 1.64 \cdot 10^3$

So the total of budgeted outlays is $1,640 billions or $1.640 trillions

Exhibit 1-57 Budget calculations.

To convert to tons, type $3.285 \cdot 10^{23} \cdot \text{kg}$ followed by =,
then double click on the last black placeholder to bring up a dialog box of units.
The dialog box is already set to mass units, so you only need to scroll down until
you get to tons. Then double click on ton to insert the new unit and close the
dialog box. Then do the same thing for the other planetary masses.

Masses in kilograms

Mercury	$3.285 \cdot 10^{23}$	$3.285 \cdot 10^{23} \cdot \text{kg} = 3.621 \bullet 10^{20}$ °ton
Venus	$4.868 \cdot 10^{24}$	$4.868 \cdot 10^{24} \cdot \text{kg} = 5.366 \bullet 10^{21}$ °ton
Earth	$5.973 \cdot 10^{24}$	$5.973 \cdot 10^{24} \cdot \text{kg} = 6.584 \bullet 10^{21}$ °ton
Mars	$6.415 \cdot 10^{23}$	$6.415 \cdot 10^{23} \cdot \text{kg} = 7.071 \bullet 10^{20}$ °ton
Jupiter	$1.899 \cdot 10^{27}$	$1.899 \cdot 10^{27} \cdot \text{kg} = 2.093 \bullet 10^{24}$ °ton
Saturn	$5.69 \cdot 10^{26}$	$5.69 \cdot 10^{26} \cdot \text{kg} = 6.272 \bullet 10^{23}$ °ton
Uranus	$8.721 \cdot 10^{25}$	$8.721 \cdot 10^{25} \cdot \text{kg} = 9.613 \bullet 10^{22}$ °ton
Neptune	$1.027 \cdot 10^{26}$	$1.027 \cdot 10^{26} \cdot \text{kg} = 1.132 \bullet 10^{23}$ °ton
Pluto	$1.493 \cdot 10^{22}$	$1.493 \cdot 10^{27} \cdot \text{kg} = 1.646 \bullet 10^{24}$ °ton
Sun	$1.99 \cdot 10^{30}$	$1.99 \cdot 10^{30} \cdot \text{kg} = 2.194 \bullet 10^{27}$ °ton

Exhibit 1-58 Mass calculation.

Energy calculation

In 1905, Albert Einstein published his now-famous equation $E = mc^2$ which showed the amount
of energy that would be produced by converting an amount of matter with mass m into energy.
The term c in the equation represents the speed of light. This calculation shows a simple example
of how Mathcad allows you to calculate with very large numbers. Mathcad also has the capability
of dealing with physical units and converting from one type of unit to another (British thermal
units to joules, for example). We will further explore that capability a little later.

We calculate the force using kilograms-meters-second and converting the 93 million mile average distance between the Sun and the Earth to meters

So the answer is in newtons

$$\frac{6.67\cdot10^{-11}\cdot\dfrac{m^2\cdot newton}{kg^2}\cdot\left(5.973\cdot10^{24}\right)\cdot kg\cdot\left(1.99\cdot10^{30}\right)\cdot kg}{\left(93000000\cdot mi\cdot1.61\cdot\dfrac{km}{mi}\cdot1000\cdot\dfrac{m}{km}\right)^2}=3.536\bullet10^{22}\quad\circ newton$$

So the force is about 3.5 x 10^22 newtons

Exhibit 1-59 Calculation of gravitational force between earth and sun.

We want to calculate how much energy can be produced from the conversion of the mass of a one-gram object to energy. Entering the numbers into Einstein's equation, we get the result shown in Exhibit 1-60.

Lorentz contraction

The shortening of moving objects is another phenomenon explained by Einstein's special theory of relativity. This shortening, known as the *Lorentz contraction*, is explored in this section by calculating some examples for low speed (Exhibit 1-61), relatively high speed (Exhibit 1-62), and a speed very close to the speed of light (Exhibit 1-63). The key thing to note is that at low speeds the contraction is very small and, for practical purposes, unnoticeable. The third example (Exhibit 1-63) shows, however, that for speeds approaching the speed of light, the Lorentz contraction becomes very significant.

For E=mc^2 with a mass of 1 gram

$$1\cdot30000000000^2=9\bullet10^{20}\quad ergs$$

or in MKS units

$$.001\cdot300000000^2=9\bullet10^{13}\quad joules$$

Exhibit 1-60 Using Einstein's equation for the equivalence of mass and energy to compute the energy equivalent of one gram.

$$l := \sqrt{1 - \frac{10^2}{300000000^2}}$$

$l = 0.999999999999999$

Exhibit 1-61 Lorentz
contraction for a case
of relatively low speed
(10 km/s).

$$l := \sqrt{1 - \frac{100000000^2}{300000000^2}}$$

$l = 0.942809041582063$

So the object contracts to about 94% of its length when at rest

Exhibit 1-62 Lorentz contraction for an object moving at
one-third the speed of light.

A Preview of Symbolic Calculations

We have begun to see how Mathcad can do some relatively simple numerical calculations in order to lay the groundwork for the truly powerful features to come. But Mathcad can also do problems where the ultimate answer is a mathematical expression rather than a number. This ability to do symbolic calculations is covered more fully later in the book, but since this is an important capability for many types of work, we will preview it in the next two problems (Exhibits 1-64 and 1-65).

$$l := \sqrt{1 - \frac{(.9 \cdot 300000000)^2}{300000000^2}}$$

$l = 0.435889894354067$

So it contracts to about 43 1/2 % of its length at rest

Exhibit 1-63 Lorentz contraction for an object moving at
nine-tenths of the speed of light.

Mathcad simplifies an expression in terms of the variable x using the Simplify command on the Symbolic menu.

$2 \cdot x + 4 \cdot x$

$6 \cdot x$ This is the simplified expression for the sum of $2 \cdot x$ and $4 \cdot x$ just as you learned in basic algebra.

Exhibit 1-64 Symbolic calculations.

Mathcad expands an algebraic expression using the "Expand Expression" command on the "symbolic" menu

$(1 + x)^{11}$

$1 + 11 \cdot x + 55 \cdot x^2 + 165 \cdot x^3 + 330 \cdot x^4 + 462 \cdot x^5 + 462 \cdot x^6 + 330 \cdot x^7 + 165 \cdot x^8 + 55 \cdot x^9 + 11 \cdot x^{10} + x^{11}$

Exhibit 1-65 Symbolic calculations.

Mortgage

$43200 loan at 7.25% in 300 months

$$i := \frac{.0725}{12}$$

$n := 300$

$PV := 43200$

$$PMT := PV \cdot \frac{i}{\left[1 - (1 + i)^{-n} \right]}$$

$PMT = 312.253$

Exhibit 1-66 A mortgage calculation.

Determine the root of a polynomial

Note that to type this equation you need to type a space after you type x^3 in order to tell Mathcad that you want to go back to the main level of the expression. Otherwise Mathcad will continue to enter your input at the exponent level. So you actually type x^3 [Space] $-2 \cdot x - .7 =$ 0   to enter this expression. You could also select the "raise to power" symbol from the palette.

$$x^3 - 2 \cdot x = -.7$$

rewrite as $x^3 - 2 \cdot x - .7 = 0$

Use Mathcad's solve capability

x := 1 <=== Guess value
$\text{root}(x^3 - 2 \cdot x - .7 , x) =$

x := 1

$$\text{root}\left(x^3 - 2 \cdot x - .7, x\right) = 1.565$$

If you select this answer and type Ctrl-F
You will see the full 15 decimal places at the bottom
of the screen
1.564506037017104
Alternatively

$$r := \text{root}\left(x^3 - 2 \cdot x - .7, x\right)$$

Then   r = 1.565   Mathcad still carries the full 15 decimal places
internally.
You could now check the answer by substituting that long number shown above into the original equation, but there is a better way that does not require you to accurately retype 15-place numbers.

Substitute r back into the original equation. This variable always carries the full precision.

$$r^3 - 2 \cdot r - .7 = 3.967 \cdot 10^{-4}$$

Approximately zero, so this is a root

There will be more on solving for roots, later in the book

Exhibit 1-67 Root of a polynomial.

Solve for a root of the following equation

cos(x)-x=.2
Rewrite the equation as   cos(x-x-.2=0
and make a guess as to the value of the root

x := 1 <=== Guess value
root(cos(x) - x - .2,x) =

$x := 1$

$root(\cos(x) - x - .2, x) = 0.616$

Using Ctrl-F we find the answer to 15 decimal places

Checking the answer by substitution

$\cos(0.6161378916856245) - .6161378916856245 - .2 = -2.149 \cdot 10^{-5}$

Approximately zero, confirming that it is a root

There will be more on solving for roots, later in the book

Exhibit 1-68 Root of an equation.

Additional Problems

Now that you have been introduced to some of Mathcad's basic capabilities, let's try a few more problems (Exhibits 1-66 to 1-75) to test our mastery of what we have covered.

Blackbody Radiation

This last problem represents a departure from what has preceded it in this chapter. We want the reader to get at least a glimpse of the power of Mathcad in a realistic problem. This is Max Planck's famous formula for calculation of the blackbody spectrum of radiation. This formula was a major step for physics at the end of the nineteenth century and laid the groundwork for the revolutionary quantum mechanics that was to come.

For this example of blackbody calculation, we use a temperature of 2.76 K (degrees Kelvin), very close to absolute zero. This makes the blackbody spectrum correspond to the radiation from the "Big Bang" which started the evolution of our universe. The temperature would have initially been extremely high, but has cooled to this very low temperature over the 15 billion years or so that the universe has been expanding. We also need to introduce two new concepts for this section.

Circumference of a circle of radius "r"

$r := 4$

$C := 2 \cdot \pi \cdot r$

$C = 25.133$

Area of a circle of radius "r"

$r := 4$

$A := \pi \cdot r^2$

$A = 50.265$

Volume of a sphere of radius "r"

$r := 4$

$V := \dfrac{4}{3} \cdot \pi \cdot r^3$

$V = 268.083$

Surface of a sphere of radius "r"

$r := 4$

$S := 4 \cdot \pi \cdot r^2$

$S = 201.062$

Exhibit 1-69 Area, volume, and circumference.

$x := 3$

$$T := \sqrt{1 + x \cdot \sqrt{1 + (x+1) \cdot \sqrt{1 + (x+2) \cdot \sqrt{1 + (x+3) \cdot \sqrt{1 + (x+4) \cdot \sqrt{1 + (x+5) \cdot \sqrt{1 + (x+6) \cdot \sqrt{1 + (x+7)}}}}}}}}$$

$T = 3.97$

The great Indian mathematician, Ramanajan, showed that when this expression of nested square root terms is carried out to infinity the right side equals "x+1"

Exhibit 1-70 Nested square roots.

Assignment of numbers to a variable

To assign a value to a variable name, type the name of the variable followed by the "assignment" operator, which is the colon ":" key on the keyboard, or the ":=" symbol on the arithmetic palette. (If you move your mouse cursor onto this operator on the palette, a small help note will appear on your screen telling you that this is the "assign value" operator). The value that you assign to the variable could be a number or a mathematical expression. Beginning with version 7, you can also assign values to a variable using the ordinary equal sign, in most cases.

The base of natural logarithms
($e = 2.718...$)

You may remember that one of the most important mathematical constants (along with pi, π) is a number known as the "base of natural logarithms" and usually referred to as e. Mathcad recognizes both π and e and knows their values. Both terms can be found on the first (arithmetic) palette. You can check this for yourself if you enter π into a Mathcad document or enter e^x and type "1" in the placeholder. Then type an equal sign, "=," immediately after. Mathcad will display the value of the constant. These concepts are used in Exhibit 1-76 in the calculation of one point on a blackbody radiation spectrum.

Basics of Saving and Printing Your Work

If you are already familiar with working on a computer, you may want to skip this section since you probably already know what to do. Otherwise, having completed some calculations, you will want to know how to preserve your work for future use.

A ball is thrown upward from the edge of a cliff at time t=0. Its vertical position then given
(in feet) by the equation   z=a*t-16*t^2
a is the initial upward speed
where z is the vertical position of the ball and the time, t, is in seconds

For t=0

$t := 0$

$a := 64$

$z := a \cdot t - 16 \cdot t^2$

$z = 0$

For t=.5

$t := .5$

$a := 64$

$z := a \cdot t - 16 \cdot t^2$

$z = 28$

For t=1

$t := 1$

$a := 64$

$z := a \cdot t - 16 \cdot t^2$

$z = 48$

For t=2

$t := 2$

$a := 64$

$z := a \cdot t - 16 \cdot t^2$

$z = 64$

For t=3

$t := 3$

$a := 64$

$z := a \cdot t - 16 \cdot t^2$

$z = 48$

Exhibit 1-71 Falling object.

For t=4

$t := 4$

$a := 64$

$z := a \cdot t - 16 \cdot t^2$

$z = 0$

For t=5

$t := 5$

$a := 64$

$z := a \cdot t - 16 \cdot t^2$

$z = -80$

In a later chapter we will see how Mathcad gives you a better way to calculate this

Exhibit 1-71 (*Continued*)

To save a document, go to the File menu and select "Save as." You will then be asked to select a name for the document. Once you type in a name, you may either click on "Save" with the mouse, or else use the "Enter" key to save the document under the name that you have chosen. For a document that you have worked on and saved previously, use "Save" on the File menu instead of "Save as."

How many miles are there in a light-year?
Light travels at 186,000 miles per second (approximately)
There are 60 seconds in a minute, 60 minutes in an hour and 24 hours in a day

$60 \cdot 60 \cdot 24 = 8.64 \cdot 10^4$ or 86,400 seconds in a day

$186000 \cdot 86400 \cdot 365.25 = 5.87 \cdot 10^{12}$

so a light-year is a little less than 6 trillion miles

Exhibit 1-72 A light-year.

How long is one billion seconds?

As we saw earlier, there are 86,400 seconds in a day, so we divide this into one billion $(1 \cdot 10^9)$

$$\frac{1 \cdot 10^9}{86400} = 1.157 \cdot 10^4 \qquad \text{or 11,574 days (approximately)}$$

$$\frac{1.157 \cdot 10^4}{365.25} = 31.677 \qquad \text{or about 31.7 years}$$

A trillion seconds is one thousand times as long (over 31,000 years)

A million seconds is $\dfrac{1 \cdot 10^6}{86400} = 11.574 \qquad$ or 11 1/2 days

Exhibit 1-73 A billion seconds.

What Is the Kinetic Energy of a 6.0 kg. Bowling Ball moving at 4.5 m/s?

Kinetic Energy=1/2 mv^2

KE=1/2 (6.0 kg) (4.5 m/s)^2

$$KE := \frac{1}{2} \cdot 6.0 \cdot 4.5^2$$

KE = 60.75   or KE=60.75 (kg m^2)/s^2   or KE=60.75 Joules

Exhibit 1-74 Bowling-ball kinetic energy.

Coulomb's law for the force between two charged particles is

F=k*m1*m2/r^2

where m1 and m2 are the two charges

k=9.0*10^9   N*m^2/C^2

and r is the distance between the charges

For this example, assume each of the charges is 1 Coulomb
and the distance between the charges is 1 meter

$$F := 9.0 \cdot 10^9 \cdot \frac{1 \cdot 1}{1^2}$$

$$F = 9 \cdot 10^9$$

So the force is 9 billion Newtons (about 2 billion pounds)
Most charged objects we would encounter in ordinary experience
have a net charge much less than a coulomb

Exhibit 1-75 Electrostatics.

Blackbody radiation for the extremely low temperature of 2.76 degrees Kelvin (the temperature of
the radiation from the Big Bang) is computed by this example.

By changing the frequency of the radiation "f" you can trace out the curve of the spectrum.

NOTE: because of the extremely small size of some of the numbers in this problem, it is necessary
to set the "zero tolerance" on the numerical format (Math menu) to 307 or some of the numbers will
be too small to register.

Exhibit 1-76 Blackbody radiation.

$t := 2.76$

$f := 1 \cdot 10^{12}$

$h := 6.625 \cdot 10^{-27}$

$c := 2.9979 \cdot 10^{10}$

$c3 := c^3$

$c3 = 2.694 \bullet 10^{31}$

$k := 1.38 \cdot 10^{-16}$

$f3 := f^3$

$a := \dfrac{h \cdot f}{k \cdot t}$

$a = 17.394$

$i := \dfrac{1}{\left(2.7182718^a - 1\right)}$

$g := 8 \cdot 3.14150 \cdot h$

$z := g \cdot i \cdot \dfrac{f3}{c3}$

$z = 0$

So for radiation at a frequency of one thousand gigahertz, the energy density of radiation is about 1.73*10^-28 ergs per cm^3 per hertz for a 2.76 degree blackbody radiator

For other frequencies:

$1 \cdot 10^8$ $3.55 \cdot 10^{-30}$

$1 \cdot 10^9$ $3.52 \cdot 10^{-28}$

$1 \cdot 10^{10}$ $3.25 \cdot 10^{-26}$

$1 \cdot 10^{11}$ $1.32 \cdot 10^{-24}$

$1 \cdot 10^{12}$ $1.73 \cdot 10^{-28}$

Exhibit 1-76 (*Continued*)

Once you have saved and closed a document, you can resume work on that document by selecting "Open" from the File menu. This will cause Mathcad to present you with a list of files that you have previously saved. Use your mouse to select the one you wish to work on, and double-click, or type the Enter key.

To print the current document, select "Print" from the File menu. Mathcad will print the document using the computer's printer. To see a preview of what the document will look like when printed, select "Print Preview" from the File menu.

2

Range Variables and Plotting

Introduction

In the first chapter, you saw how to do some basic operations including simple editing and computing. You now know enough to try some of your own work. In this chapter you will greatly expand your Mathcad expertise by learning how Mathcad uses range variables to give you more computing and plotting power.

Precision of Displayed Results

Mathcad carries numbers internally as a 1-digit characteristic and a 15-digit mantissa (16 digits total). But Mathcad's developers knew that for many types of work, it would be much clearer to have your displayed results rounded to a lower precision. So Mathcad shows you three decimal places, but allows you to select some other value. To set the precision of the results as you want them to be displayed on your computer screen or printed, go to the Format menu and select "Number." You will then see the "Number Format" dialog box, which allows you to set the desired degree of precision for your displayed results. Exhibit 2-1 is a screen print of this dialog box.

At the top of the dialog box, Mathcad allows you to select the number system for your work; decimal numbers are the default setting and the one you will use most of the time, but you could choose to use "Hex" (hexadecimal) or "Octal" where appropriate for your work. You may also choose to use j instead of the default i for the square root of -1 in the imaginary parts of (complex) numbers.

In the section of the dialog box with the heading "Precision," Mathcad gives you boxes where you can type in your own choices for the degree of precision with which answers are displayed. The "Displayed Precision" choice allows you to select an integer between 0 and 15. This number will determine how many decimal places Mathcad shows in computed results. The default value is 3. Understand that this choice affects only the number as displayed. Mathcad continues to use 15 decimals internally, regardless of what you select here.

Exhibit 2-2 gives an example of how this works. Some digits of the mathematical constant

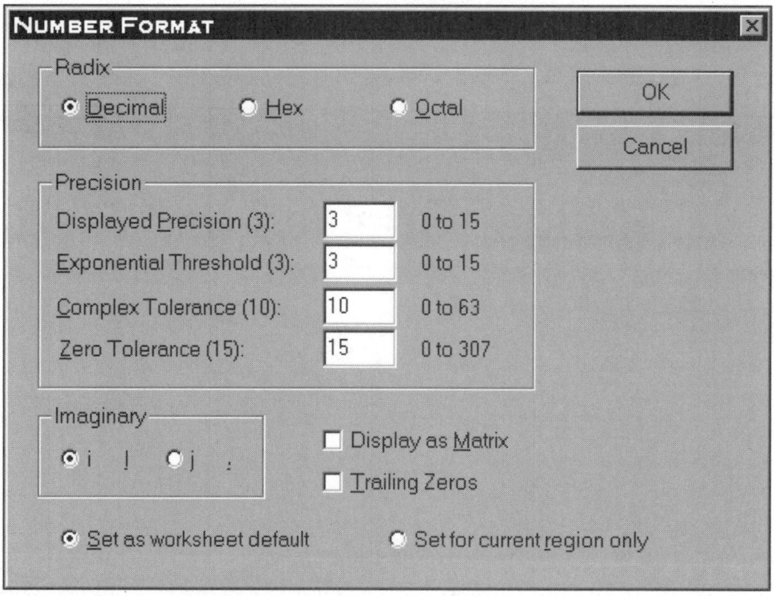

Exhibit 2-1 Number Format dialog box.

π have been typed into a Mathcad screen. You recall, of course, that π is an unending number. It has an infinite number of digits. For practical purposes, Mathcad, like other math programs, uses a truncated version of the complete number. To show such a numerical approximation of π, we then first entered a π into the Mathcad document by clicking on the symbol on the arithmetic palette. We then caused Mathcad to evaluate numerical value by entering the Greek letter pi followed by an equal sign. This caused Mathcad to display the (truncated) numerical value of π to four significant figures (value of 3 plus three decimal places).

We now do the same operation again, and then we reset the display precision, using the "Number Format" dialog box, to 15 decimals. The scope of this precision definition is determined by choosing "Set as Worksheet Default" or "Set for Current Region Only" at the bottom of the Number Format dialog box. For this example "Set for Current Region Only" is selected, meaning that only this calculation is governed by this selection. Mathcad now shows the value of π to be 3 plus 15 decimal places (16 digits total) for the new calculation; the displayed precision of the first calculation is still unchanged at the default precision.

$\pi = 3.142$

$\pi = 3.141592653589793$

Exhibit 2-2 Numerical approximation of π.

$\pi = 3.141592653589793$

$\pi = 3.141592653589793$

Exhibit 2-3 Numerical approximation of π.

If we select Set as Worksheet Default instead of Set for Current Region Only and repeat the same operations, all computed numbers are shown with 15 decimal places as in Exhibit 2-3. So we see that Set as Worksheet Default sets the precision for the whole Mathcad document, but Set for Current Region Only affects only the current calculation.

The "Exponential Threshold" setting allows you to set the level at which Mathcad begins to show the computed results in exponential (powers-of-10) notation. In Exhibit 2-4, we see the results of setting the Exponential Threshold to 2 and then to 15. The default is 3.

The "Trailing Zeros" choice allows you to control the number of digits that are displayed to the right of the decimal point. If you check this box, Mathcad will use the number chosen for the displayed precision as the number of decimal digits to show in displayed results. For example, in Exhibit 2-5, we have selected 5 to be the "Displayed Precision," and we have checked the Trailing Zeros box. Consequently, when we displayed the numerical values of the constants π and e, they are displayed to five decimal places. We then assigned a numerical value to a newly defined variable a—by using the assignment operator and typing in a number. Then, when we require Mathcad to display the numerical value of a by typing "a" followed by an equal sign, the numerical result is shown to five decimal places.

If this explanation leaves you a little confused about the difference between displayed precision and trailing zeros, consider the following example. Create a variable (we will call it "a") which is equal to 222222.33333333 and set the exponential threshold to 15, to prevent some numbers being displayed in powers-of-10 notation, which will confuse the point we are illustrating. Then evaluate the variable a 6 times and compare your results to the following table. (All these examples have Exponential Threshold set to 15.)

Displayed precision	Trailing Zeros not checked	Trailing Zeros checked
3	222222.33333333	222222.333
8	222222.33333333	222222.33333333
10	222222.33333333	222222.3333333300

In all the examples above, notice that the numbers to the left of the decimal point are the same. But the Displayed Precision choice, and the choice of checking, or not checking, the Trailing Zeros box affects how the decimal part of the number is shown. As expected, in the bottom example, with Trailing Zeros checked, Mathcad displays two zeros to fill in the missing numbers. Try it yourself to see how it works.

The "Display as Matrix" box is for work which generates answers with more than nine rows or columns. By default, Mathcad generates a scrolling table containing your output in order

$a := 123456.7890456$

$a = 1.235 \cdot 10^5$ Exponential Threshold set to 2

$a \cdot a = 15241578761.45$ Exponential Threshold set to 15

$\dfrac{a}{100000000} = 1.235 \cdot 10^{-3}$ Exponential Threshold set to 2, Displayed Precision set to 3

$\dfrac{a}{100000000} = 0.001234567890456$ Exponential Threshold set to 15, Displayed Precision set to 15

Exhibit 2-4 Use of Exponential Threshold setting.

___

to make a more compact presentation. In such an instance, if you prefer to have the results displayed in a long vector or matrix that would show all the results without scrolling, you should check the Display as Matrix box. This may result in a long vector or matrix that extends to subsequent pages, but will enable you to see and print your entire results. The importance of this selection will become more apparent later in this chapter when we begin to use range variables to generate and plot multiple values in a calculation.

The "Complex Tolerance" and "Zero Tolerance" choices will be explained later.

Accuracy and Precision

We have now seen how to set the precision with which numbers are displayed by Mathcad. It is important to keep in mind that the precision with which results are displayed is not the same thing as the accuracy of those results. It is quite possible for you to set display parameters that show many more decimal places in your results than are warranted by the accuracy of your data. It is also possible to do calculations to many more decimal places than are warranted

___

Displayed precision is 5, Trailing zeros box is checked

$\pi = 3.14159$

$e = 2.71828$

$a := 273.25689127$ (typed in)

$a = 273.25689$ (variable "a" evaluated by Mathcad)

Exhibit 2-5 Using Trailing Zeros option.

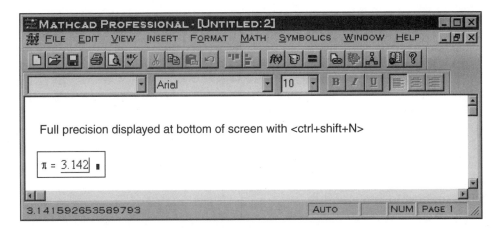

Exhibit 2-6 Screen print showing full-precision number.

by your data. For example, assume that you have measured the radius of a circle to be 6.0 cm. You estimate that your measurement is accurate to one decimal place (a total of two significant places). If you now calculate the circumference of a circle with this radius by multiplying this by 2 times π, Mathcad will calculate the circumference to 16 figures. If you set the displayed precision to show all 16 of these digits, you will have an answer that, to the naïve observer, appears to be very precise, yet only two of the digits in the answer are actually meaningful—since the radius measurement had only two significant figures. A complete discussion of this topic is beyond the scope of this book, but you should be aware that the displayed precision of an answer may not be a true indication of its accuracy.

There are times when Mathcad shows a result of a calculation and you want to know the number in its full precision, without changing its displayed precision. To see the number in full (as Mathcad represents it internally), depress the "Ctrl," "Shift," and "N" keys simultaneously. The full-precision number will be shown at the bottom of the Mathcad screen. See Exhibit 2-6 for a screen print illustrating this. In this exhibit you see that the numerical value of π has been shown to the default precision (three decimal places) on the screen. By typing "Ctrl+Shift+N," you can see all 16 digits of precision just below the horizontal scroll bar.

How Mathcad Calculates a Worksheet

When you are learning how to set up your problem in a Mathcad worksheet, it is important that you have some understanding of how Mathcad uses your worksheet to do your calculations. Mathcad starts at the top of the worksheet and works its way across the page to the right and then down, just as you would do in reading a document. As it scans through your document, Mathcad looks for places where you have assigned values to variables and then uses those values for the variables the next time it does calculations. Mathcad expects to see a value assigned for the variable, before you try to use it in your work. The value assignment could be on a line above the line where the variable is used, or it could be to the left on the same line. If a variable is assigned a value in more than one place, Mathcad uses the last value assigned before the calculation. For example, if you assign a value and do a calculation, you could then assign a new value and do a new calculation, and Mathcad will use the last value assigned each time it calculates. If Mathcad does not find a value assignment before the variable is used, it will display an error message that will tell you that you have not properly defined the variable. Note that when you do a symbolic calculation, this does not apply. In symbolic expressions, the variables do not have specific numerical values (unless you assign them such values). The answers are mathematical expressions (formulas), not specific numbers. As a result, Mathcad deals with symbolic expressions differently, as we shall see in a later chapter.

Using Range Variables in Mathcad

The problems we have dealt with up to this point have used Mathcad's ability to calculate a single answer to a specific mathematical expression. But Mathcad's real power comes from its ability to handle problems in which you repeatedly change the value of a variable and do many recalculations. These answers then become multiple data points that can be compiled in a table or plotted on a graph to show the shape of the curve. Mathcad makes it easy to do problems such as this by employing a concept known as the *range variable,* a variable that takes a number of values over a specified range. In Mathcad a range variable can be used to tell the Mathcad program what values to give the independent variable in your calculation. A range variable acts somewhat like a computer programming statement that tells the computer to "do [some operation] for values of 1 to 10." In other words, Mathcad takes the initial value of the range variable, calculates the value of the expression and then moves to the next value, does another calculation, and continues until it has used all values of the range variable. Range variables greatly increase the power of Mathcad by giving you the ability to set up a calculation procedure and then have Mathcad do many iterations of it.

A word of warning is in order at this point. Some people find the use of range variables to be a little tricky at first. But you should have no difficulty using them if you work through this section and try some examples for yourself.

To define a range variable, type the name of a variable, then type a colon ":" or use the assignment operator on the arithmetic palette, just as you would to make any assignment of a value to a variable. The next step (this is what makes it a range variable instead of an ordinary variable) is to type the starting number, typically a 0 or a 1. Then type a comma and enter the second value (starting value plus one step). Finally type a semicolon. Mathcad then enters some periods (an ellipsis) to tell you that some values are coming after the second value, but they are not shown. It also puts a placeholder after the ellipsis to tell you that it is looking for you to enter another value—the ending value. Once you enter this final value, if you omit the second value Mathcad assumes that you want it to use the initial value, then add 1 to that initial value, then keep stepping up the value in increments of 1 until it reaches the final value. In essence, you are writing a very simple program to tell Mathcad what number to start with and how many times to iterate the calculation, but you do not have to learn a programming language to do this. Mathcad continues to show the results in normal mathematical notation. Once you get used to using range variables, you will find them to be a simple key to getting Mathcad to do difficult calculations for you. Furthermore, if you decide to change some parameters and recalculate, Mathcad will do it for you quickly.

In Exhibit 2-7, we see how a range variable is defined. It follows the procedure described above. The first range variable, called "t," in Exhibit 2-7 starts at 0 and steps through values 1, 2, 3, and so on until it reaches 9. To confirm the values, after the assignment was made, we typed the name of the variable followed by an equal sign "=," and Mathcad displayed the values. A second variable called "var" has also been defined, with values ranging from 1 to 15.

Mathcad will allow you to define a range variable with steps of 2 or 3 or anything instead of 1. But when you use it in your calculations, it will have zeros for the missing steps as in Exhibit 2-8.

You may be wondering if it is possible for the range to extend into negative numbers. Exhibit 2-9 gives an example of this. As you can see, although Mathcad allows you to define such a variable, it will not allow you to use it as a subscript for a variable unless the array origin option is set to the first range value. The "a_v" is in red, to indicate that you have done something wrong, and when you click on the "a_v," a box appears with the message "Value of subscript or superscript is too big (or too small) for this array." Since range variables are commonly used as subscripts, this limits the usefulness of ones that extend into negative numbers. We will see how to cope with this later.

Will Mathcad allow nonintegral steps for range variables? Again the answer is "yes, you can define such a variable, but no, you cannot use it as an index." See Exhibit 2-10 for an example. Here we used steps of 0.2. If you click on the subscript, Mathcad tells you "This must be an integer. The expression you used appears to have a fractional part." If you want more detail about what is wrong, use the F1 key to summon up additional help.

Mathcad will allow you to define a range variable in descending order as shown in Exhibit 2-11. You will see that when you evaluate the variable (have Mathcad print its values), the values will be in descending order. However, if you use the variable as a subscript for a vector in a calculation, it will give results in ascending rather than descending order, because the sequence of subscripts is always ascending.

What can you do with a range variable?

You can use it to add a sequence of numbers as in Exhibit 2-12. Here we have defined a range variable and then used the sigma operator "Σ" on the calculus palette to sum the values in the range variable. We clicked on the sigma operator from the calculus palette and typed the starting and ending values in the placeholders below and above the sigma. The expression we wanted to sum was simply the various values taken by "t," the range variable. So we typed "t" in the place-holder to the right of the sigma.

A more interesting use for a range variable is shown in Exhibit 2-13. Here, after the range variable "n" was defined, it was used as an index (a subscript) for a vector "A." Vector "A" in this example is an exponential function, and the range variable tells Mathcad to evaluate the function for values of 1 through 9. Notice that Mathcad also inserts a value for zero when the "A" function is evaluated, although the range variable started with 1. Mathcad expects the first value in a vector to be the zeroth. So even though Mathcad will allow you to define a

Define a range variable

t := 0 .. 9

Evaluate t

t

0
1
2
3
4
5
6
7
8
9

Define another range variable

var := 1 .. 15

var

1
2
3
4
5
6
7
8
9
10
11
12
13
14
15

Exhibit 2-7

Define a range variable

n := 1 , 3 .. 30

Evaluate n

n

1
3
5
7
9
11
13
15
17
19
21
23
25
27
29

Define $b_n := n$

$b =$

	0
0	0
1	1
2	0
3	3
4	0
5	5
6	0
7	7
8	0
9	9
10	0
11	11
12	0
13	13
14	0
15	15

Exhibit 2-8

$$t := 1, 1.2 .. 7$$

t

1
1.2
1.4
1.6
1.8
2
2.2
2.4
2.6
2.8
3
3.2
3.4
3.6
3.8
4
4.2
4.4
4.6
4.8
5
5.2
5.4
5.6
5.8
6
6.2
6.4
6.6
6.8
7

$$v := -10 .. 10$$

v

-10
-9
-8
-7
-6
-5
-4
-3
-2
-1
0
1
2
3
4
5
6
7
8
9
10

$$a_v := v$$

$$a_t := t$$

$$t := 12 .. 1$$

t

12
11
10
9
8
7
6
5
4
3
2
1

$$a_t := t$$

$$a =$$

	0
0	0
1	1
2	2
3	3
4	4
5	5
6	6
7	7
8	8
9	9
10	10
11	11
12	12

Exhibit 2-9   **Exhibit 2-10**   **Exhibit 2-11**

$t := 1 .. 30$

t

1
2
3
4
5
6
7
8
9
10
11
12
13
14
15
16
17
18
19
20
21
22
23
24
25
26
27
28
29
30

$$\sum_{t=1}^{30} t = 465$$

$n := 1 .. 9$

$A_n := e^n$

$$A = \begin{array}{c|c} & 0 \\ \hline 0 & 0 \\ 1 & 2.718 \\ 2 & 7.389 \\ 3 & 20.086 \\ 4 & 54.598 \\ 5 & 148.413 \\ 6 & 403.429 \\ 7 & 1.097 \cdot 10^3 \\ 8 & 2.981 \cdot 10^3 \\ 9 & 8.103 \cdot 10^3 \end{array}$$

$n := 0 .. 9$

$A_n := e^n$

$$A = \begin{array}{c|c} & 0 \\ \hline 0 & 1 \\ 1 & 2.718 \\ 2 & 7.389 \\ 3 & 20.086 \\ 4 & 54.598 \\ 5 & 148.413 \\ 6 & 403.429 \\ 7 & 1.097 \cdot 10^3 \\ 8 & 2.981 \cdot 10^3 \\ 9 & 8.103 \cdot 10^3 \end{array}$$

$t := 1 .. 60$

$$S := 1 + \sum_t \frac{1}{t!}$$

$S = 2.718281828459$

$e = 2.718281828459$

Exhibit 2-12   **Exhibit 2-13**   **Exhibit 2-14**   **Exhibit 2-15**

range variable starting with a value other than zero, it will put in zeros for the missing calculations. In most cases this will not cause a problem, but you should be aware of it. Exhibit 2-14 is the same as Exhibit 2-13 except that the range variable starts at 0 (not 1). The first calculation now yields 1, since "e" raised to the zero power is 1.

Exhibit 2-15 shows how a range variable can be used to calculate the approximate value of an infinite series. Although such a series has an infinite number of terms, if the series converges to an answer quickly enough, it may be possible to compute a good approximation of its value by using only a relatively small number of terms. Here the (infinite) series expansion for "e" is calculated and displayed to 11-decimal-place precision. To test the accuracy of the approximation we have compared the calculated value (from the series) to the known value of "e" (the value Mathcad has stored internally) to the same number of decimal places. A variable ranging from 1 to 60 is adequate to compute the series to 11-decimal-place accuracy, as you can see in this exhibit.

Exhibit 2-16 shows how Mathcad can be used to compute an approximation to the famous function known as the *Riemann zeta* function. The function is of great interest in advanced mathematical work. We are not going to attempt an explanation of its significance to mathematicians here, but it does provide a nice example of the power of Mathcad. The zeta function can be defined by the infinite series

$$\text{Zeta} = 1 + (1/2)^2 + (1/3)^2 + (1/4)^2 + \cdots \text{etc.}$$

The zeta series converges very slowly, so you need to compute a lot of terms to get a good approximation to its value. You could do the calculation by explicitly writing out many terms of the series. But Mathcad provides more efficient ways to do such calculations, as shown in Exhibit 2-16.

Will Mathcad allow us to use two range variables in the same calculation?

The next three examples show that the answer is "yes" and give some illustrations of how they can be used.

In Exhibit 2-17 we have defined two range variables called "i" and "j." We then used "i" and "j" as subscripts for a variable "S"—defined as the sum of "i" and "j." By evaluating "S," we see the results of the computations. Note that Mathcad has inserted zeros in the top row and the leftmost column of the matrix of values for "S."

In Exhibit 2-18 we again defined range variables "i" and "j." This time we use them as exponents and again evaluate "S" to produce the matrix of results. In this example, because of the size of the numbers in the answer, Mathcad has cut off some of the columns to the right and produced a scrolling display. To see the missing results, click on the matrix and use the scrolling arrow that appears to scroll to the missing results.

Also, if you prefer to see all the answers in one list (for example, in order to print them), you will remember that the Math Format dialog box discussed earlier had a check box that caused Mathcad to display results as a matrix instead of in a scrolling list.

For Exhibit 2-19, we use "i" as the argument of an exponential function "j" as the argument of a sine function. Again, the computed results produce a scrolling horizontal display. In this example we see that although Mathcad did not label the axes of the matrix, the values along

$$Z(k) := \sum_{n=1}^{10000} \frac{1}{n^k}$$

$Z(4) = 1.0823232337$

$Z(5) = 1.0369277551$

$Z(6) = 1.0173430620$

$Z(7) = 1.0083492774$

$Z(8) = 1.0040773562$

$Z(9) = 1.0020083928$

$Z(10) = 1.0009945751$

If you want Mathcad to give you more accuracy, you can write the sum to infinity and then on the Symbolic menu do a floating point evaluation. Mathcad will ask the number of places of precision. For 20 places, the result for Z(4) is as follows.

$$\sum_{n=1}^{\infty} \frac{1}{n^4}$$

1.0823232337111381916

For 50 places

$$\sum_{n=1}^{\infty} \frac{1}{n^4}$$

$1.0823232337111381915160036965411679027747509519187$

Exhibit 2-16

$i := 1 .. 10$

$j := 1 .. 10$

$S_{i,j} := i + j$

$S =$

	0	1	2	3	4	5	6	7	8	9	10
0	0	0	0	0	0	0	0	0	0	0	0
1	0	2	3	4	5	6	7	8	9	10	11
2	0	3	4	5	6	7	8	9	10	11	12
3	0	4	5	6	7	8	9	10	11	12	13
4	0	5	6	7	8	9	10	11	12	13	14
5	0	6	7	8	9	10	11	12	13	14	15
6	0	7	8	9	10	11	12	13	14	15	16
7	0	8	9	10	11	12	13	14	15	16	17
8	0	9	10	11	12	13	14	15	16	17	18
9	0	10	11	12	13	14	15	16	17	18	19
10	0	11	12	13	14	15	16	17	18	19	20

Exhibit 2-17

the top are the "j" values and the ones along the left side are the "i" values. For example, we see that the number under the column heading "3" and in the row labeled "7" is 154.757. If you compute

$$\sin(3) = .141120$$
$$e^7 = 1096.63315$$

by multiplication, you get .141120 * 1096.63315 = 154.757.

Exhibit 2-20 shows a cosine function multiplied by an exponential. A range variable "j" goes from 0 to 200, producing an oscillation, such as you might get from the solution to a differential equation, that dies out after a few cycles.

Exhibits 2-21, 2-22, and 2-23 are closely related examples that show how Mathcad handles

- The sine of an argument that is squared, such as $\sin(\chi^2)$
- The sine of an argument multiplied by the sine of the argument $\sin(\chi) \bullet \sin(\chi)$
- The sine of a sine $\sin(\sin(\chi))$

Note that the second example above, which we would normally write as $\sin^2(\chi)$, is one of the few instances where Mathcad deviates from standard math notation.

$i := 1 .. 10$

$j := 1 .. 10$

$S_{i,j} := e^{i+j}$

$S =$

	0	1	2	3	4	5
0	0	0	0	0	0	0
1	0	7.389	20.086	54.598	148.413	403.429
2	0	20.086	54.598	148.413	403.429	$1.097 \cdot 10^3$
3	0	54.598	148.413	403.429	$1.097 \cdot 10^3$	$2.981 \cdot 10^3$
4	0	148.413	403.429	$1.097 \cdot 10^3$	$2.981 \cdot 10^3$	$8.103 \cdot 10^3$
5	0	403.429	$1.097 \cdot 10^3$	$2.981 \cdot 10^3$	$8.103 \cdot 10^3$	$2.203 \cdot 10^4$
6	0	$1.097 \cdot 10^3$	$2.981 \cdot 10^3$	$8.103 \cdot 10^3$	$2.203 \cdot 10^4$	$5.987 \cdot 10^4$
7	0	$2.981 \cdot 10^3$	$8.103 \cdot 10^3$	$2.203 \cdot 10^4$	$5.987 \cdot 10^4$	$1.628 \cdot 10^5$
8	0	$8.103 \cdot 10^3$	$2.203 \cdot 10^4$	$5.987 \cdot 10^4$	$1.628 \cdot 10^5$	$4.424 \cdot 10^5$
9	0	$2.203 \cdot 10^4$	$5.987 \cdot 10^4$	$1.628 \cdot 10^5$	$4.424 \cdot 10^5$	$1.203 \cdot 10^6$
10	0	$5.987 \cdot 10^4$	$1.628 \cdot 10^5$	$4.424 \cdot 10^5$	$1.203 \cdot 10^6$	$3.269 \cdot 10^6$

Exhibit 2-18

Exhibit 2-24 demonstrates how Mathcad can be used to calculate and graphically display values for a calculation. Here we have used a range variable from 1 to 212 to represent Fahrenheit temperatures and have used the familiar Fahrenheit-to-Celsius conversion equation to convert these Fahrenheit temperatures to Celsius. This is a relatively simple calculation, but it displays some of the power of Mathcad in that it is able to relieve you of the drudgery of these many calculations. Notice that because of the large number of values in this calculation, Mathcad did not display the entire list of answers for variable "F" (the Fahrenheit temperatures) or for "C" (the Celsius temperatures). It showed only some of the values, but gave you a scrolling list. To see another section of the list, you simply click the mouse anywhere on the list, and a scroll bar will appear. Notice that the corresponding values of the index (range variable t) are shown to the left of the values of "F" and "C" when we evaluated them. As a result, you can readily see that 14°F is equivalent to -10°C. Finally, you will notice that we have provided a Mathcad-produced graph of the results to show the relationship between the two temperature scales. To create

$i := 1 .. 10$

$j := 1 .. 10$

$S_{i,j} := e^i \cdot \sin(j)$

$S =$

	0	1	2	3	4
0	0	0	0	0	0
1	0	2.287	2.472	0.384	-2.057
2	0	6.218	6.719	1.043	-5.592
3	0	16.901	18.264	2.834	-15.201
4	0	45.943	49.646	7.705	-41.32
5	0	124.885	134.952	20.944	-112.319
6	0	339.474	366.837	56.932	-305.316
7	0	922.785	997.166	154.757	-829.935
8	0	$2.508 \cdot 10^3$	$2.711 \cdot 10^3$	420.673	$-2.256 \cdot 10^3$
9	0	$6.819 \cdot 10^3$	$7.368 \cdot 10^3$	$1.144 \cdot 10^3$	$-6.132 \cdot 10^3$
10	0	$1.853 \cdot 10^4$	$2.003 \cdot 10^4$	$3.108 \cdot 10^3$	$-1.667 \cdot 10^4$

Exhibit 2-19

this graph, we simply chose the Insert menu and "Graph" and "X-Y plot." We then entered the names of the variables we wished to plot in the placeholders along the x axis and the y axis. Note that you can also call up a graph from the keyboard by typing "@" or pressing the Shift Key and the number 2 simultaneously.

Exhibit 2-25 is a companion to Exhibit 2-24, in that it shows the other side of the same relationship. Here, Celsius temperatures are converted to Fahrenheit ones. Here you can see that 59°F is equivalent to 15°C.

You may think there is something vaguely unsatisfying about these exhibits because they do not extend to negative temperatures. It would be interesting to see the point (−40°) at which the two temperature scales give the same temperature. Furthermore, from a practical standpoint, to be useful in relatively cold climates, any temperature-related function has to extend into the negative temperatures.

This raises a significant point that you need to understand about how Mathcad uses range variables. Mathcad does not allow negative range variables to be used as indexes in a calculation unless the array origin option is set to the first range value. For many of the problems you work on, this will not be a limitation, but there are times when you will need to define a new

$$t := 1 .. 100$$

$$S_t := \sin\left(\frac{t^2}{20^2}\right)$$

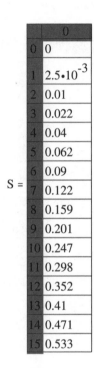

	0
0	0
1	$2.5 \cdot 10^{-3}$
2	0.01
3	0.022
4	0.04
5	0.062
6	0.09
7	0.122
8	0.159
9	0.201
10	0.247
11	0.298
12	0.352
13	0.41
14	0.471
15	0.533

$S =$

$$j := 0 .. 200$$

$$S_j := \exp\left(-\frac{j}{20}\right)^2 \cdot \cos(j)$$

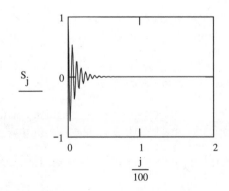

Exhibit 2-20

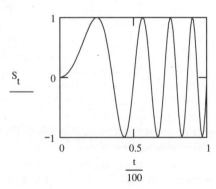

Exhibit 2-21

$i := 1 .. 100$

$s_i := \sin\left(\dfrac{i}{20}\right)^2$

	0
0	0
1	$2.498 \cdot 10^{-3}$
2	$9.967 \cdot 10^{-3}$
3	0.022
4	0.039
5	0.061
6	0.087
7	0.118
8	0.152
9	0.189
10	0.23
11	0.273
12	0.319
13	0.366
14	0.415
15	0.465

$s =$

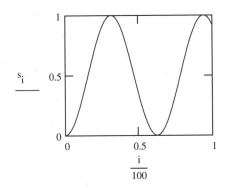

Exhibit 2-22

$i := 1 .. 100$

$s_i := \sin\left(\sin\left(\dfrac{i}{20}\right)\right)$

	0
0	0
1	0.05
2	0.1
3	0.149
4	0.197
5	0.245
6	0.291
7	0.336
8	0.38
9	0.421
10	0.461
11	0.499
12	0.535
13	0.569
14	0.601
15	0.63

$s =$

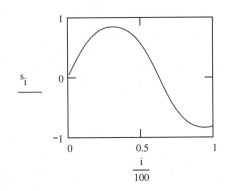

Exhibit 2-23

$t := 1 .. 212$

$F_t := t$

$C_t := \dfrac{5}{9} \cdot \left(F_t - 32\right)$

$t := 1 .. 100$

$C_t := t$

$F_t := \dfrac{9}{5} \cdot C_t + 32$

$F =$

	0
0	0
1	1
2	2
3	3
4	4
5	5
6	6
7	7
8	8
9	9
10	10
11	11
12	12
13	13
14	14
15	15

$C =$

	0
0	0
1	-17.222
2	-16.667
3	-16.111
4	-15.556
5	-15
6	-14.444
7	-13.889
8	-13.333
9	-12.778
10	-12.222
11	-11.667
12	-11.111
13	-10.556
14	-10
15	-9.444

$F =$

	0
0	0
1	33.8
2	35.6
3	37.4
4	39.2
5	41
6	42.8
7	44.6
8	46.4
9	48.2
10	50
11	51.8
12	53.6
13	55.4
14	57.2
15	59

$C =$

	0
0	0
1	1
2	2
3	3
4	4
5	5
6	6
7	7
8	8
9	9
10	10
11	11
12	12
13	13
14	14
15	15

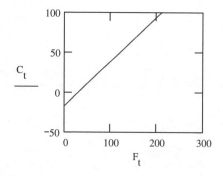

Exhibit 2-24

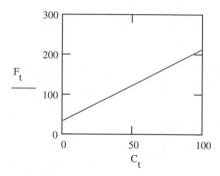

Exhibit 2-25

variable or use another strategy for doing that problem. In Exhibits 2-26 and 2-27 we show one approach to such a problem. In these examples we have defined a new variable: the range variable less 100. This allows us to use a positive range variable and still calculate negative temperatures.

Exhibit 2-28 shows a result that you might get from a critically damped oscillator. Notice that the range variable took on values from 1 to 300 in order to get enough data points to create a smooth graph. How many values do you need? You will find from trial and error or from experience with the type of work you do the most what is appropriate. We have found that for some types of curves that include oscillations, it is often necessary to use range variables that take relatively large numbers of values (for example, 300 or 400).

Exhibit 2-29 shows how to compute Fibonnaci numbers from the defining formula and using a range variable.

Exhibit 2-30 is an example of using a range variable to compute some integral powers of e. The results are then graphed, first on a regular x-y plot and then on a scale where the y axis is logarithmic. The first plot shows a curve with an increasingly sharply rising curve. This second plot straightens a sharply rising plot into a straight line. To create a plot with a logarithmic, go to the Insert menu and choose "Graph," then "X-Y Plot." After creating the graph, use the Format menu to select "Graph," and "X-Y Plot." Mathcad will show you the formatting choices for the x axis and y axis, as well as which are currently selected. To make the y axis logarithmic, click on "Log Scale" under "Y-axis."

In Exhibit 2-31 we use a range variable to generate calculations of the periods of revolution for satellites at various altitudes above the surface of the earth. For this example, the range variable goes from 1 (1 mi above the surface of the earth) to 100 mi above the surface. The 7950 in the equation is the earth's diameter. It is divided by 2, to convert it to the earth's radius, and the value of the range variable "h" is added, because we want to calculate the height above the earth's surface. The factor of 5280 converts the height from miles to feet. The denominator is a number based on the strength of the earth's gravitation. From this we calculate the time "P" (in seconds) it takes a satellite at the given distance from the earth to go through one orbit. The final part of our calculation converts the time for one revolution around the orbit into minutes. The calculations ignore any effect of air resistance.

Astronomy: The Moons of Uranus

To illustrate how to use some of the graphing capabilities of Mathcad, we will use some published data from a spacecraft that explored the outer planets of the solar system a decade ago.

When the Voyager 2 spacecraft flew past Uranus in 1986, it returned a wealth of data about the planet and its 15 moons (10 of which were not known prior to the flyby). The following chart shows the data to be used:

Name (or designation)	x-Data Distance from center of Uranus, thousands of kilometers	y-Data orbital period, hours
1986 U7	49.7	8.0
1986 U8	53.8	9.0
1986 U9	59.2	10.4
1986 U3	61.8	11.1
1986 U6	62.7	11.4
1986 U2	64.6	11.8
1986 U1	66.1	12.3
1986 U4	69.9	13.4
1986 U5	76.3	14.9
1985 U1	86.0	18.3
Miranda	129.9	33.9
Ariel	190.9	60.5
Umbriel	266.0	99.5
Titania	436.3	208.9
Oberon	583.4	323.1

To plot these in Mathcad, we create a range variable. We will call the range variable "n" (see Exhibit 2-32), and it will have 15 values, for the 15 pairs of x and y data pairs shown in the table above.

The distances from the center of Uranus will be the x-axis values and will be entered into a vector called "a." To create this vector, use the "Insert" menu and select "Matrix." But first, create a variable "a" and when Mathcad gives you a placeholder to enter the value for "a," go to the Insert menu and create a matrix. Your matrix (really a vector) will need to have 15 rows and 1 column. When Mathcad asks you how many rows and columns you want in your newly created matrix, select 15 rows and 1 column. Now click on the first placeholder in your new vector and enter the first number. Then use the "Tab" key to move to the next placeholder and enter the second number. Continue this process until you have filled in all the x values in this vector.

The times for the moons to circle their orbits one time will be the y-axis values and will be entered into a vector called "b." Follow the same procedure used above to create this "b" vector, or the other method described on the exhibit.

To produce the graph, we use the Mathcad Insert menu and select "Graph." Alternatively, we could type "@." This causes Mathcad to place a template for a graph at the cursor position.

The template has several placeholders (small black rectangles, where we can enter data for Mathcad to act on). We are going to need only the one in the middle of the x axis and the one in the middle of the y axis at present. Use your mouse to click on the placeholder in the middle of the x axis and type "a_n" to tell Mathcad to plot the elements of the "a" vector along the x axis. Do a similar operation to have Mathcad plot "b_n" along the y axis. Remember to use the "[" Key to create the subscript.

In order to have our conversion work for sub-zero temperatures we could introduce a
new variable equal to the range variable minus 100. A better approach is shown below.
Type ORIGIN := -100. This makes the column head of the display-as-matrix table equal -100.
The C temperature corresponding to any F temperature in the gray column is now shown in the
white column.

ORIGIN := -100

$t := -100 .. 100$

$F_t := t$

$C_t := \dfrac{5}{9} \cdot (F_t - 32)$

$C =$

	-100
-47	-43.889
-46	-43.333
-45	-42.778
-44	-42.222
-43	-41.667
-42	-41.111
-41	-40.556
-40	-40
-39	-39.444
-38	-38.889
-37	-38.333
-36	-37.778
-35	-37.222
-34	-36.667
-33	-36.111
-32	-35.556

Notice that when the Fahrenheit
temperature is -40, the Celsius
temperature is also -40, as it
should be

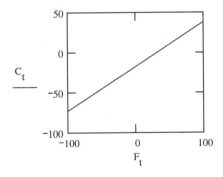

Exhibit 2-26

For this Celsius to Fahrenheit conversion we reset the origin as in the previous example.

ORIGIN := -100

t := -100 .. 100

$C_t := t$

$F_t := \dfrac{9}{5} \cdot C_t + 32$

$$
F = \begin{array}{|c|c|}
\hline
 & -100 \\
\hline
-100 & -148 \\
\hline
-99 & -146.2 \\
\hline
-98 & -144.4 \\
\hline
-97 & -142.6 \\
\hline
-96 & -140.8 \\
\hline
-95 & -139 \\
\hline
-94 & -137.2 \\
\hline
-93 & -135.4 \\
\hline
-92 & -133.6 \\
\hline
-91 & -131.8 \\
\hline
-90 & -130 \\
\hline
-89 & -128.2 \\
\hline
-88 & -126.4 \\
\hline
-87 & -124.6 \\
\hline
-86 & -122.8 \\
\hline
-85 & -121 \\
\hline
\end{array}
$$

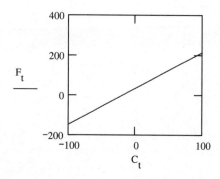

Exhibit 2-27

$k := 1 .. 300$

$$y_k := e^{\left(-\frac{k}{50}\right)} \cdot \sin\left(\frac{k}{10}\right)$$

$y =$

	0
0	0
1	0.098
2	0.191
3	0.278
4	0.359
5	0.434
6	0.501
7	0.56
8	0.611
9	0.654
10	0.689
11	0.715
12	0.733
13	0.743
14	0.745
15	0.739

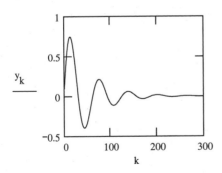

Exhibit 2-28

Fibonacci Numbers

ORIGIN := 0

$a_0 := 1$

$a_1 := 1$

n := 2 .. 50

$a_n := a_{n-1} + a_{n-2}$

$$a = \begin{array}{c|c}
 & 0 \\
\hline
0 & 1 \\
1 & 1 \\
2 & 2 \\
3 & 3 \\
4 & 5 \\
5 & 8 \\
6 & 13 \\
7 & 21 \\
8 & 34 \\
9 & 55 \\
10 & 89 \\
11 & 144 \\
12 & 233 \\
13 & 377 \\
14 & 610 \\
15 & 987 \\
\end{array}$$

Note that in this example we are using the range variable to drive true iterations in which the new values of the sequence are being generated based on values generated earlier. We are not just plugging the index value into a formula, we are using the sequence to feed back into itself.

Exhibit 2-29

$n := 1 .. 10$

$a_n := e^n$

	0
0	0
1	2.718
2	7.389
3	20.086
4	54.598
5	148.413
6	403.429
7	$1.097 \cdot 10^3$
8	$2.981 \cdot 10^3$
9	$8.103 \cdot 10^3$
10	$2.203 \cdot 10^4$

$a =$

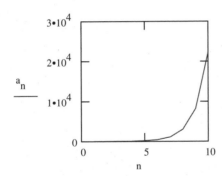

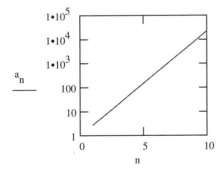

Exhibit 2-30

For heights of 1 mile to 100 miles above the Earth's surface, compute the
period of revolution of a satellite--assuming no air resistance. Show the
period "P" in seconds and in minutes.

$h := 1 .. 100$

$$P_h := 2 \cdot \pi \cdot \sqrt{\frac{\left[\left[\left(\frac{7950}{2}\right) + h\right] \cdot 5280\right]^3}{1.407 \cdot 10^{16}}}$$

	0
0	0
1	5095.107
2	5097.029
3	5098.951
4	5100.874
5	5102.797
6	5104.721
7	5106.644
8	5108.568
9	5110.492
10	5112.416
11	5114.341
12	5116.265
13	5118.19
14	5120.115
15	5122.041

$P =$

	0
0	0
1	84.918
2	84.95
3	84.983
4	85.015
5	85.047
6	85.079
7	85.111
8	85.143
9	85.175
10	85.207
11	85.239
12	85.271
13	85.303
14	85.335
15	85.367

$\dfrac{P}{60} =$

Exhibit 2-31

n := 0.. 14

$$a := \begin{bmatrix} 49.7 \\ 53.8 \\ 59.2 \\ 61.8 \\ 62.7 \\ 64.6 \\ 66.1 \\ 69.9 \\ 76.3 \\ 86.0 \\ 129.9 \\ 190.9 \\ 266.0 \\ 436.3 \\ 583.4 \end{bmatrix}$$

$b_n :=$

8.0
9.0
10.4
11.1
11.4
11.8
12.3
13.4
14.9
18.3
33.9
60.5
99.5
208.9
323.1

Tip: The method of entering the data used here for a is the straightforward approach. A nice technique some people prefer uses an input table. Having used the direct method for a, we use the input table method to enter data in b.

It works as follows:

After defining your range variable as above,
type   b[n := 8.0, 9.0, 10.4, etc.
Each comma you type gives you a new line in
the input table. Be careful to use the proper bracket.

Example:

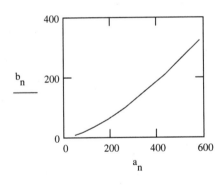

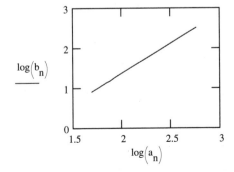

Exhibit 2-32

Having seen how easy it is to plot data in Mathcad, we will do another plot where the logarithms of the vector elements are calculated before the data are plotted. Just create another graph, as before, but type "log(a_n)" instead of "a_n" and "log(b_n)" instead of "b_n" on the axes. You will see that the logarithms have transformed a curved plot into a straight line.

If we calculate the slope of the straight line, we find that it equals ³⁄₂ or 1.5:

$$\text{Slope} = \frac{\Delta y}{\Delta x}$$

$$\frac{\log(323.1) - \log(8.0)}{\log(583.4) - \log(49.7)} = \frac{1.61}{1.07} = 1.5$$

This is a nice confirmation of Kepler's third law of planetary motion, although Kepler had no knowledge of the existence of these moons:

> The squares of the periods of revolution of the planets about the sun [here read "moons about their planet"] are proportional to the cubes of their mean distances from the sun [here read "planet"].

Incidentally, another way to produce the second graph would have been to define two new vectors which are the logarithms of the "a" and "b" vectors, then plot these vectors.

In Exhibit 2-33 we compute the wavelengths of some lines in the Balmer series of the spectrum of hydrogen. The simple formula

$$109678 \cdot \left(\frac{1}{m^2} - \frac{1}{n^2} \right)$$

gives a good approximation to many of the emission lines in the hydrogen spectrum. If $m = 2$, the lines are called the *Balmer series*. The first line is for $n = 3$, and is a bright-red line at the wavelength of 656.3 nm. As n increases in value, the lines get fainter and closer together, eventually converging at 364.6 nm. Beyond this is a region of continuous spectrum. Other values of m produce additional series that lie outside the range of visible light.

In Exhibit 2-34 we investigate an old rule of thumb used by people working with mathematics in the days before Mathcad. It was said that for relatively small angles, the value of the angle (in radians) is approximately equal to the value of its sine and also to the value of its tangent. Therefore, problems could often be simplified by using the angle in place of the sine or tangent if the problem could be restricted to small angles. Here we use a range value to compute some values and determine the truth of this rule. We compute three variables: "A," which is the radian equivalent of 1°, 2°, and so on; "SIN," which is the sine of the angle; and "TAN," which is the tangent of the angle. We see that for angles up to about 6° (approximately 0.087 radians), the angle (expressed in radians), the sine of the angle, and the tangent of the angle are equal to three decimal places—good enough for many rough calculations.

In Exhibit 2-35 we calculate the trajectory of an object thrown vertically upward at an initial speed of 16 ft/s. This is similar to a problem we did in Chap. 1, but now we can make use of a range variable to step through values of time. To get a fine-grained picture of the

$n := 3 .. 1000$

$$S_n := 109678 \cdot \left(\frac{1}{2^2} - \frac{1}{n^2} \right)$$

$$\lambda_n := \frac{1}{S_n}$$

$\lambda =$

	0
0	0
1	0
2	0
3	$6.565 \cdot 10^{-5}$
4	$4.863 \cdot 10^{-5}$
5	$4.342 \cdot 10^{-5}$
6	$4.103 \cdot 10^{-5}$
7	$3.971 \cdot 10^{-5}$
8	$3.89 \cdot 10^{-5}$
9	$3.836 \cdot 10^{-5}$
10	$3.799 \cdot 10^{-5}$
11	$3.772 \cdot 10^{-5}$
12	$3.751 \cdot 10^{-5}$

Exhibit 2-33

movement, we have divided each increment of the range variable by 10 to show the movement one-tenth of a second at a given time. So our 30 increments represent 3 s. We see, increment by increment, how the object rises (as its initial upward velocity is offset by the downward acceleration due to gravity) and then how it begins to fall.

In Exhibit 2-36 we examine (again using range variables) a simple exponential function and see how it can be separated into a real part, which becomes a cosine series; and an imaginary part, which becomes a sine series. First, we create an exponential function with an imaginary argument (e^{it}) where i is the square root of minus 1 and t is the angle. It is well known that the real part of such a function

$$\frac{e^{it} + e^{-it}}{2}$$

is the cosine function and the imaginary part

$$\frac{e^{it} - e^{-it}}{2}$$

is the sine function. Here we demonstrate this using Mathcad and also show Mathcad's ability to handle such trigonometric and exponential functions. In later chapters we will give more examples of working with trigonometric and exponential functions. Here we also use Mathcad's built-in "real" and "imaginary" operators to compute the real and imaginary parts of the exponential function and compare them to the results calculated as described above and to the values of cosine and sine.

In Exhibit 2-37 we examine two sine waves of slightly different frequencies. This recalls

---

$n := 1 .. 12$

$$A_n := \frac{n}{360} \cdot 2 \cdot \pi$$

When you want to suppress the index (to make a simpler display of your output), you can use an output table by typing A[n= instead of A= , sin(A[n)= and cos(A[n)= as shown below.

A_n	$\sin\left(A_n\right)$	$\tan\left(A_n\right)$
0.017	0.017	0.017
0.035	0.035	0.035
0.052	0.052	0.052
0.07	0.07	0.07
0.087	0.087	0.087
0.105	0.105	0.105
0.122	0.122	0.123
0.14	0.139	0.141
0.157	0.156	0.158
0.175	0.174	0.176
0.192	0.191	0.194
0.209	0.208	0.213

Exhibit 2-34

$n := 1 .. 30$

$t_n := \dfrac{n}{10}$

The basic expression is:

$$y_n := 16 \cdot t_n - \frac{1}{2} \cdot 9.8 \cdot \left(t_n\right)^2$$

	0
0	0
1	1.551
2	3.004
3	4.359
4	5.616
5	6.775
6	7.836
7	8.799
8	9.664
9	10.431
10	11.1
11	11.671
12	12.144
13	12.519
14	12.796
15	12.975

$y =$

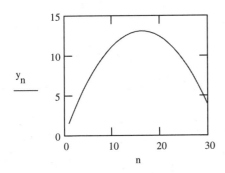

Exhibit 2-35

$\cos(\theta)$ = real part of $e^{i \cdot \theta}$

$t := 0 .. 10$

$C_t := e^{i \cdot t}$

$C =$

	0
0	1
1	0.54+0.841i
2	-0.416+0.909i
3	-0.99+0.141i
4	-0.654-0.757i
5	0.284-0.959i
6	0.96-0.279i
7	0.754+0.657i
8	-0.146+0.989i
9	-0.911+0.412i
10	-0.839-0.544i

$Re(C) =$

	0
0	1
1	0.54
2	-0.416
3	-0.99
4	-0.654
5	0.284
6	0.96
7	0.754
8	-0.146
9	-0.911
10	-0.839

$c_t := \dfrac{e^{i \cdot t} + e^{-i \cdot t}}{2}$

$c =$

	0
0	1
1	0.54
2	-0.416
3	-0.99
4	-0.654
5	0.284
6	0.96
7	0.754
8	-0.146
9	-0.911
10	-0.839

$\cos(t)$

1
0.54
-0.416
-0.99
-0.654
0.284
0.96
0.754
-0.146
-0.911
-0.839

Exhibit 2-36

$\sin(\theta)=$ the imaginary part

$t := 0..\ 10$

$S_t := e^{i \cdot t}$

	0
0	0
1	0.841i
2	0.909i
3	0.141i
4	-0.757i
5	-0.959i
6	-0.279i
7	0.657i
8	0.989i
9	0.412i
10	-0.544i

$s_t := \dfrac{e^{i \cdot t} - e^{-i \cdot t}}{2} \qquad s =$

$Im(C) =$

	0
0	0
1	0.841
2	0.909
3	0.141
4	-0.757
5	-0.959
6	-0.279
7	0.657
8	0.989
9	0.412
10	-0.544

$\sin(t)$

0
0.841
0.909
0.141
-0.757
-0.959
-0.279
0.657
0.989
0.412
-0.544

Exhibit 2-36 (*Continued*)

the result from basic physics or from music classes that when two sounds of slightly different frequencies are sounded together, as when two tuning forks or two piano keys of slightly different frequencies are sounded together, a new sound is created. Here we use Mathcad to add two waves of slightly different frequency and illustrate the production of such "beats."

In Exhibit 2-38 we see an equation that represents the first two terms of a series that is the solution to a heat-flow problem. The problem is to find the heat distribution in a uniform

$i := 1 .. 1000$

$S_i := \sin\left(\dfrac{i}{100} \cdot 2 \cdot 3.14159\right)$

$T_i := \sin\left(\dfrac{i}{100} \cdot 1.1 \cdot 2 \cdot 3.14159\right)$

$Z_i := S_i + T_i$

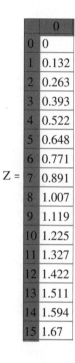

	0
0	0
1	0.132
2	0.263
3	0.393
4	0.522
5	0.648
6	0.771
7	0.891
8	1.007
9	1.119
10	1.225
11	1.327
12	1.422
13	1.511
14	1.594
15	1.67

$Z =$

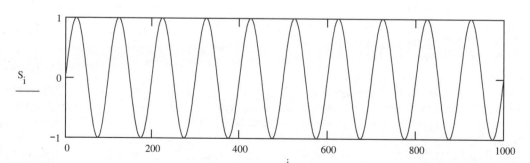

Exhibit 2-37

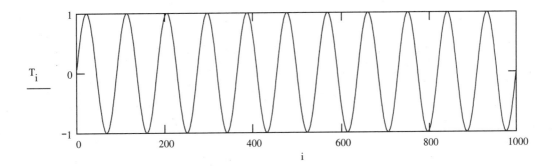

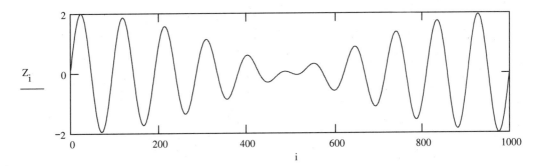

Exhibit 2-37 (*Continued*)

horizontal bar of length "l" which is positioned along the x axis. Initially the temperature distribution in the bar is given by

$$f(x) = \begin{cases} x & \text{for} \quad 0 < x < (\frac{1}{2}) \, 1 \\ 1 - x & \text{for} \quad (\frac{1}{2}) \, 1 < x < 1 \end{cases}$$

By facilitating the calculation of the heat distribution in the bar and plotting the results, Mathcad allows you to understand what is happening over time. The first graph (Exhibit 2-38) shows the results for time 0.1 assuming that the length of the bar is equal to 100. The results for time 1000 (Exhibit 2-39) and time 10,000 (Exhibit 2-40) are also shown. The temperature becomes small throughout the bar as the time becomes large.

Exhibit 2-41 is a problem that one of the authors was asked to help with by a coworker a few years ago. The coworker was trying to prove

$$\sin^6 \theta + \cos^6 \theta = 1 - 3 \cdot \sin^2 \theta \cdot \cos^2 \theta$$

$x := 1 .. 100$

$c := 1$

$1 := 100$

$t := .1$

$$U_x := 4 \cdot \frac{1}{\pi^2} \cdot \left[\sin\left(\pi \cdot \frac{x}{1}\right) \cdot e^{-\left(c \cdot \frac{\pi}{1}\right)^2 \cdot t} - \left(\frac{1}{9}\right) \cdot \sin\left(\frac{3 \cdot \pi \cdot x}{1}\right) \cdot e^{-\left(3 \cdot c \cdot \frac{\pi}{1}\right)^2 \cdot t} \right]$$

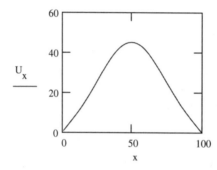

Exhibit 2-38

---

but was having trouble doing so. He asked for help in demonstrating the validity of this equation for some range of values. Exhibit 2-41 shows the Mathcad demonstration of this.

Polar Plots

In Exhibits 2-42 and 2-43 we show how you can use Mathcad to produce polar plots. For polar plots we first specify a value for "N" (the number of points to be plotted.) We have used 50 here, but if you use a larger number, your plot will be somewhat smoother. Next, you must define the values of "θ" (theta) over which your plot is to drawn. This is done in the same way as you create a range variable, but you specify the initial value of "θ," the next value, and (following the ellipses) the final value of "θ." For polar graphs, the final value will often (but not always) be 2 times "π" (pi). Finally, you must specify the definition of the function in terms of "r" the radius and "θ" the angle. For example, in Exhibit 2-42, the radius is defined as a constant times the sine of 3 times "θ." After you have completed these definitions, you create the polar plot by clicking your mouse at the position on the Mathcad screen where the plot

$x := 1 .. 100$

$c := 1$

$l := 100$

$t := 1000$

$$U_x := 4 \cdot \frac{1}{\pi^2} \cdot \left[\sin\left(\pi \cdot \frac{x}{l}\right) \cdot e^{-\left(c \cdot \frac{\pi}{l}\right)^2 \cdot t} - \left(\frac{1}{9}\right) \cdot \sin\left(\frac{3 \cdot \pi \cdot x}{l}\right) \cdot e^{-\left(3 \cdot c \cdot \frac{\pi}{l}\right)^2 \cdot t} \right]$$

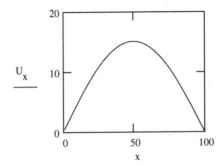

Exhibit 2-39

x := 1 .. 100

c := 1

l := 100

t := 10000

$$U_x := 4 \cdot \frac{1}{\pi^2} \cdot \left[\sin\left(\pi \cdot \frac{x}{l}\right) \cdot e^{-\left(c \cdot \frac{\pi}{l}\right)^2 \cdot t} - \left(\frac{1}{9}\right) \cdot \sin\left(\frac{3 \cdot \pi \cdot x}{l}\right) \cdot e^{-\left(3 \cdot c \cdot \frac{\pi}{l}\right)^2 \cdot t} \right]$$

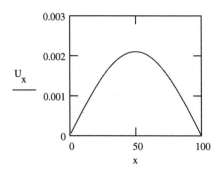

Exhibit 2-40

$t := 0 .. 100$

$\theta_t := \dfrac{t}{5}$

$c_t := \sin\left(\theta_t\right)^6 + \cos\left(\theta_t\right)^6$

$d_t := 1 - 3 \cdot \cos\left(\theta_t\right)^2 \cdot \sin\left(\theta_t\right)^2$

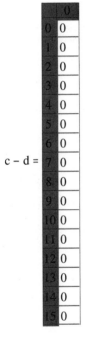

$c - d =$

	0
0	0
1	0
2	0
3	0
4	0
5	0
6	0
7	0
8	0
9	0
10	0
11	0
12	0
13	0
14	0
15	0

This shows that the difference is essentially 0 and so confirms the conjectured relationship.

A quicker check is to take the length of this difference vector. If the length is very small, all the elements must be nearly 0.

In fact

$$\left| c - d \right| = 1.727 \bullet 10^{-15}$$

Exhibit 2-41

should be located. Mathcad will place a red cursor (in the shape of a cross) at this location. Now select "Polar Plot" from the "Graph" submenu of the Insert menu. Mathcad will insert the skeleton of a polar plot with placeholders for you to insert information about what you want to be plotted. You have already defined the values for "θ," so you should insert the Greek letter "θ" using the Greek alphabet palette in the bottom placeholder, and the function name for "r" in the left placeholder. Mathcad will now draw the plot as shown in the exhibit.

The Mathcad Treasury, available from MathSoft Inc. for an extra charge, has a number of

N := 50

a := 1.2

$$\Theta := 0, 2 \cdot \frac{\pi}{N} .. 2 \cdot \pi$$

$$r(\Theta) := a \cdot \sin(\Theta)$$

N := 50

$$\Theta := 0, 2 \cdot \frac{\pi}{N} .. 2 \cdot \pi$$

$$r(\Theta) := \left(\cos\left(\frac{1}{2} \cdot \Theta \right) \right)^2$$

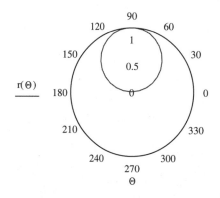

Exhibit 2-42

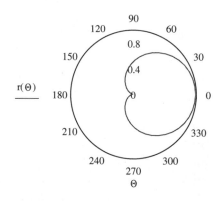

Exhibit 2-43

other polar plots, including nice examples of the spiral of Archimedes and the lemniscate including animated examples. If you expect to do polar plots, you should look at these examples.

Refreshing Your Screen

You will sometimes find that after editing your work, the screen becomes unclear or confused (some parts of numbers or equations may be cut off or difficult to read). If this happens, you should go to the View menu and select "Refresh." Mathcad will then clarify the view by redrawing the equations on the screen. In the authors' experience, this happens infrequently, usually following periods of extensive editing, but when it does happen, you should be aware that it can be easily corrected.

3

Integration and Complex Numbers

Introduction

In the first two chapters, we saw how to use some of the basic functions of Mathcad that you need to master in order to get started. In the next several chapters we explore some bread-and-butter capabilities of Mathcad that will be especially useful for getting your work done. In this chapter we concentrate on two capabilities of Mathcad that have great practical usefulness for many of the types of problems you may want to solve. This chapter shows how Mathcad helps you with integration of functions and with complex numbers. We should note that, try as we might to segregate features of Mathcad into separate chapters and sections in order to deal with them one at a time, there is inevitably some overlap. Some topics that we would like to keep in their own sections insist on insinuating themselves into other sections. So, for example, we cannot truly separate integration from complex numbers or differentiation from integration because many problems involve more than one of these topics.

Because of its usefulness in many types of practical calculations, the calculus, invented independently by Newton and Leibniz, is a key mathematical tool in many types of problems you might work with.

Mathcad can help you work with the calculus, especially with such practical tools as solving integration problems. This chapter is intended to help you get started using Mathcad with this type of work and to suggest examples of problems that Mathcad can help you with. The remainder of the chapter will help you work with complex numbers, another essential tool for many types of work in physics, mathematics, and engineering.

Here the authors have tried to select varied examples of integrals that illustrate various aspects of Mathcad's integration computational power.

Integrals

Integration is one of the most powerful and practical of tools in any math worker's toolkit. Integration can be used to calculate areas, volumes, lengths, average values of functions, surface areas, and many other useful types of results. Since integrals are so valuable to many

$$\int_{\blacksquare}^{\blacksquare} \blacksquare \, d\blacksquare$$

$$\int \blacksquare \, d\blacksquare$$

Exhibit 3-1 Definite integral.

Exhibit 3-2 Indefinite integral.

types of work, the fact that they are often difficult or impossible to compute analytically make the ease with which Mathcad can do numerical (and often symbolic) integration of great value in solving problems.

A few words of review are in order here. You will recall that integrals can be classified into one of the two categories:

- Definite integrals (integrals that have specified limits, as in Exhibit 3-1)
- Indefinite integrals (an example of which is seen in Exhibit 3-2)

The value of a properly set-up definite integral can always be calculated numerically—although you can conceivably set up integrals that do not have calculable values because the limits of integration do not make sense or the value of the integral is infinite.

You also probably remember that some (not all) integrals that can be solved by numerical methods can also be solved symbolically—to yield an answer that is a formula, rather than a number. When a symbolic or analytic solution is obtained to an integral, an infinite number of other solutions are also available by adding any arbitrary amount, as when e^x is integrated to $e^x + C$, where C is an arbitrary constant. Mathcad has powerful methods for finding answers to integrals by both numerical and symbolic techniques.

The primary type of integral to be dealt with in this chapter is the definite integral, and we are concerned here mainly with numerical methods of solving them. A later chapter deals in more depth with some of Mathcad's powerful methods for dealing with many types of symbolic calculations (not just integrals), but you will soon see that many integrals can be solved by both numerical and symbolic methods. As a result, there will be some discussion and examples of symbolic methods in this chapter, although that is not the primary topic to be discussed here. Mathcad's integra-

$$a := \int_{1}^{1.5} \frac{1}{\sqrt{2 \cdot x - x^2}} \, dx$$

$$a = 0.524$$

Exhibit 3-3

$$\int_{1}^{1.5} \frac{1}{\sqrt{2 \cdot x - x^2}} \, dx = 0.524$$

Exhibit 3-4

$$\int_{0}^{2 \cdot \pi} \cos(\theta)^2 \, d\theta = 3.142$$

Exhibit 3-5

$$TOL := .0000000001$$

$$b := \int_0^1 \frac{1}{1+x^2} dx$$

$$b = 0.78539816$$

$$\int_0^2 \int_0^x y \, dy \, dx = 1.333$$

$$4 \cdot b = 3.14159265$$

Exhibit 3-6   **Exhibit 3-7**

---

tion capabilities use standard mathematical notation. They are easy to apply to your work using templates and placeholders such as you have already seen.

Now assume that you have come to a point in your work where you need to solve an integral. Turn on your computer, start up Mathcad, and get ready to work with integrals. Let's assume that your first integral is a definite integral for which you require a numerical answer. You start by clicking on the "Definite Integral" sign on the Calculus palette. The template gives you placeholders for the upper and lower limits, the expression to be integrated, and the variable of integration (after the "d"). (See Exhibit 3-3.)

In the first example, a variable "a" is defined as being equal to the integral by using the colon ":" or the "Define = sign" on the palette. So once the integral values have been typed into their placeholders (the expression to be integrated, the variable with respect to which the integral is to be computed, and the limits of integration—for definite integrals), Mathcad calculates the value of the integral and sets "a" equal to that value. To see the result, just evaluate "a" (i.e., type "a" followed by an equal sign). Typing the equal sign tells Mathcad to display the current value of the variable "a," which we have defined to be equal to the value of the integral.

Exhibit 3-4 shows the same integral as shown in Exhibit 3-1, but Mathcad lets you enter it another way. The main difference is that we have dispensed with the step of setting the integral equal to a variable—so to get the computed value of the integral, we just type an equal sign "=" after the integral.

Exhibit 3-5 is another example of an integral that has been done in the same way as the preceding example (Exhibit 3-1).

Mathcad can also do multiple integrals (see Exhibit 3-6). The procedure is the same except that you produce 2, 3, or n number of integral signs (by clicking on the palette) instead of only one. Of course you must be careful to get the limits of integration and $dx \, dy \, dz$, etc. in the right order to make certain that the integrations are performed in the right sequence.

At the beginning of this example (see Exhibit 3-7), there is a "TOL" followed by a colon ":".

Numerically

$$\int_0^1 x \cdot \ln(x)\, dx = -0.25$$

Numerically

$$\int_{-1}^0 x \cdot \sqrt{x + 1}\, dx = -0.267$$

x := 0.3

Symbolically

$$\int_0^1 x \cdot \ln(x)\, dx$$

$$\frac{-1}{4}$$

Symbolically

$$\int_{-1}^0 x \cdot \sqrt{x + 1}\, dx$$

$$C(x) := \int_0^x \cos\left(\frac{\pi}{2} \cdot t^2\right) dt$$

$$\frac{-4}{15} = -0.267$$

The equal sign was typed here to show the decimal result for comparison to the numerical calculation.

C(x) = 0.299401

Exhibit 3-8 Fresnel integral.

Exhibit 3-9

Exhibit 3-10

You will remember that the colon tells Mathcad that you are about to define something. (You could also use the ":=" on the palette.) You are defining the tolerance level for the calculation. The smaller the number, the more precise the calculation will be (and the longer it will take to complete—although for simple integrals, you will not notice much difference in the calculation time). So here we have set the tolerance to 10^{-10}, instead of the default value of .001, because we want to get a more precise result. Once we have calculated the value of the integral, we multiply it by 4. We do this because this is a special integral, which has the value of $\frac{1}{4}\,\pi$. By multiplying by 4, we have calculated the value of π to eight decimal places.

Notice that in order to see all eight decimal places, you must set the number of places by going to the Format menu and selecting "Number." Then, on the line "Displayed Precision" you can select the number of decimal places you want displayed, and whether this choice is for all numbers in the document (Set as Worksheet Default) or just this number (Set for Current Region Only).

Exhibit 3-8 is an example of Fresnel's integral, which is important in the analysis of diffraction in optics. The value of this integral cannot be written as an analytic expression, but must be solved by numerical integration as done in this exhibit.

Some integrals can be solved both numerically and symbolically. Many integrals in this chapter have been done both ways. The definite integral in Exhibit 3-9 is an integral involving natural logarithms. We have already seen how to have Mathcad evaluate a definite integral numerically. To evaluate a definite integral symbolically, set it up as you would if you were going to have Mathcad evaluate it numerically, using the definite-integral symbol from the calculus palette. Then select the integral expression to be evaluated, by clicking on it with the mouse and (if need be) using the spacebar to expand the selection to the whole expression. Make sure the vertical part of the cursor is to the right and encloses the entire expression. Then select "Evaluate Symbolically" from the Symbolics menu. Mathcad will display the symbolic expression for the value of the integral directly below the selection box by default.

Numerically

$$\int_{0}^{\frac{\pi}{2}} \sqrt{1 + \cos(2 \cdot x)} \, dx = 1.414$$

Symbolically

$$\int_{0}^{\frac{\pi}{2}} \sqrt{1 + \cos(2 \cdot x)} \, dx$$

$$\int_{0}^{\infty} \ln\left(1 - e^{-x}\right) dx$$

$$\sqrt{2} = 1.414$$

$$\frac{-1}{6} \cdot \pi^2$$

Exhibit 3-11 **Exhibit 3-12**

You could also use the symbolic equal sign (a right-pointing arrow) found on the "Symbolic and Boolean" palette—or the keyboard equivalent "Control" and "plus" (Ctrl+). In this case, Mathcad will show the result to the right of the original expression.

To symbolically calculate an indefinite integral, follow the same steps outlined above, except choose the symbol for an indefinite integral (the one without integration limits) from the calculus palette instead of the definite-integral symbol.

Exhibit 3-10 is an example of an integral involving a square root.

In Exhibit 3-11 we have another example of integrals that can be evaluated either numerically or symbolically. In this example, the integrand involves both a square root and a trigonometric function.

In Exhibit 3-12 Mathcad evaluates an integrand with an exponential and a natural logarithm that produces a nice, simple result when integrated from 0 to infinity (∞).

Exhibits 3-13 and 3-14 demonstrate one of the many uses for a definite integral by showing how it can be used to compute the average value of a function over some range. The average is computed by integrating the function over the limits of integration and dividing by the difference between the two limits. This gives the average value of the function over the given range.

Exhibit 3-15 demonstrates a well-known property of definite integrals which is that reversing the limits of the integral reverses the sign of the answer. After computing the integral, we switched the limits of integration and recalculated the value of the integral. As you can see, the sign of the answer changes.

Exhibit 3-16 involves an integration of an exponential function with an absolute value in its argument. No symbolic solution is available for this integral.

Average Value of Function between x=1 and x=4

Average Value of Function

$$\frac{1}{4-1} \cdot \int_{1}^{4} x \cdot e^{\left(x^2\right)} dx = 1481017.969$$

$$\frac{1}{e^e - e} \cdot \int_{e}^{e^e} \frac{1}{x} dx = 0.138$$

Symbolically

Symbolically

$$\frac{1}{4-1} \cdot \int_{1}^{4} x \cdot e^{\left(x^2\right)} dx$$

$$\frac{1}{e^e - e} \cdot \int_{e}^{e^e} \frac{1}{x} dx$$

$$\frac{1}{6} \cdot \exp(16) - \frac{1}{6} \cdot \exp(1) = 1481017.967$$

$$\frac{1}{(\exp(e) - e)} \cdot (\exp(1) - 1) = 0.138$$

Exhibit 3-13

Exhibit 3-14

A variety of examples follow (Exhibits 3-17 to 3-34), showing how Mathcad solves integration problems.

Area by integration

So far we have been more interested in the solutions to these integral problems than in any practical applications they may have. The next few problems will explore some of the uses of

Reversing the limits of integration changes the sign

$$A := \int_{1}^{2} e^x dx$$

Numerically

$$\exp(2) - \exp(1)$$

$$A := \int_{-1}^{0} e^{|x|} dx$$

$$A1 := \int_{2}^{1} e^x dx$$

$$A = 1.718$$

$$-\exp(2) + \exp(1)$$

A symbolic solution is not available for this integral

Exhibit 3-15

Exhibit 3-16

Numerically

Numerically

$$\int_2^3 \frac{x^2 - 3 \cdot x + 4}{(x-1) \cdot (x+1)} dx = 0.542419$$

$$\int_{.1}^1 \frac{x^2 + 1}{\left(x^3 + 3 \cdot x + 3\right)} dx = 0.250562$$

Symbolically

Symbolically

$$\int_2^3 \frac{x^2 - 3 \cdot x + 4}{(x-1) \cdot (x+1)} dx$$

$$\int_{.1}^1 \frac{x^2 + 1}{\left(x^3 + 3 \cdot x + 3\right)} dx$$

$-4 \cdot \ln(4) + 1 + \ln(2) + 4 \cdot \ln(3) = 0.542419$

$\frac{1}{3} \cdot \ln(7) - .39807515095701865366 = 0.250562$

Exhibit 3-17

Exhibit 3-18

these solutions. Exhibit 3-35 is an example of an integral whose solution represents an area of the xy plane. This is a common use for integration, and before electronic methods of calculation of integrals were available, various ingenious graphical and mechanical methods of determining approximate measures of area were devised. In particular, the curve represents a plot of a probability function. The integral in this example is the area under the curve from 0 to 0.6, and it is also a probability. We will study this further in another problem.

Volume by integration

In Exhibit 3-36 we compute the volume of a sphere of radius 2 as generated by revolving (about the y axis) the area, bounded by the y axis and half of the circle $x^2 + y^2 = 4$.

Line integrals

In Exhibit 3-37 we compute the line integral $2xy\ dx + (x^2 + y^2)dy$, where the path for the integral is the parabola $y^2 = x$ between (0,0) and (1,−1).

Integration of the sun's spectrum

In this example (Exhibit 3-38) we have an expression (based on Planck's radiation formula) for the amount of power generated by each square meter of the sun's surface per unit of wavelength. The example shows how to integrate this formula over the visible wavelengths to calculate the power in visible light.

This integral cannot be done numerically because one of the limits is infinite, though you could use a large number in place of the infinite limit and approximate the value. Here we do it symbolically and using Mathcad's limit function.

Symbolically

$$\int_{1}^{\infty} \frac{1}{x^2}\,dx$$

1

Using the limit function

$$\lim_{t \to \infty} \int_{1}^{t} \frac{1}{x^2}\,dx$$

1

Exhibit 3-19

Integration of the normal distribution

In Exhibit 3-39 we determine the probability of an occurrence using the normal distribution. The distribution shown here has a mean of zero and a standard deviation of 1. Instead of using a table of values of the normal distribution, we determine the value by having Mathcad integrate the normal distribution function, first for one standard deviation from the mean— then for two standard deviations. Since the normal distribution is symmetric, and since we are only calculating with one tail of it, we double the result. We see, in our calculations, that probability of an item sampled from a normally distributed universe being within two standard deviations of the mean is about 95 percent. The probability of it being within three standard deviations is more than 99 percent.

Spiral problem 1: length of a reel of tape

In this problem (Exhibit 3-40), we consider a spool on which magnetic tape is wound. The radius of the spool is ½ in, and the outer edge of the spool of tape is an inch from the axis of the spool. The tape has a thickness of ¹⁄₁₀₀₀ in. The problem is to determine the length of the spool of tape by computing the integral shown in the exhibit.

Numerically

$i := 1 .. 10$

$$A_i := \int_0^i e^{-x} dx$$

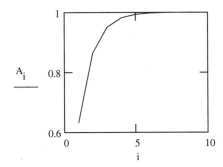

$A =$

	0
0	0
1	0.632
2	0.865
3	0.95
4	0.982
5	0.993
6	0.998
7	0.999
8	1
9	1
10	1

Exhibit 3-20

We want the answer to be in inches, so we use Mathcad's capability for working with units here. Once you gain familiarity with this part of Mathcad, you can throw away your conversion tables and let Mathcad do the work of converting between various types of units. Mathcad, by default, uses SI units (meters-kilograms-seconds-amperes-degrees kelvin-candelas-moles), but you can change the default setting by selecting "Options" from the Math menu, and then

Since this is an indefinite integral, no numerical solution exists.

$$\int \frac{\ln(x)}{x} dx$$

$$\frac{1}{2} \cdot \ln(x)^2$$

Exhibit 3-21

---

selecting the "Unit System" tab. For this problem, U.S. customary units were used, because we wanted to work with inches and feet. To create a number with an attached set of units, type the multiplication sign (*) directly after the number and follow it with the appropriate abbreviation for the unit of your choice. For example, in this problem, we defined a variable "t" to be .001 and then typed "*in" immediately after the number—as though we were multiplying the number by "in" because we want the answer in inches. You can use the placeholder that appears after a number when you evaluate a variable, to convert from one unit to a different unit. For example, if the answer is in inches and you type "ft" in the placeholder,

---

Numerically

$$\int_{\frac{\pi}{2}}^{\pi} \int_{0}^{\sqrt{2 \cdot y}} x \cdot \sin\left(y^2\right) dx\, dy = 0.451$$

Symbolically

$$\int_{\frac{\pi}{2}}^{\pi} \int_{0}^{\sqrt{2 \cdot y}} x \cdot \sin\left(y^2\right) dx\, dy$$

$$\frac{-1}{2} \cdot \cos\left(\pi^2\right) = 0.451$$

Exhibit 3-22

Numerically

$$\int_{0}^{1} \cos(x) \cdot e^{-\int_{0}^{x} \cos(t)\, dt}\, dx = 0.569$$

Symbolically

$$\int_{0}^{1} \cos(x) \cdot e^{-\int_{0}^{x} \cos(t)\, dt}\, dx$$

$$-\exp(-\sin(1)) + 1 = 0.569$$

Exhibit 3-23

Numerically

$$\int_0^{\frac{\pi}{2}} \sin(x)^3 \, dx = 0.667$$

Numerically

$$\int_1^{e^2} \ln(x^2) \, dx = 16.778$$

Numerically

$$\int_0^{\frac{\pi}{2}} \left(1 - \frac{x^2}{2!} + \frac{x^4}{4!} - \frac{x^6}{6!} + \frac{x^8}{8!}\right) \cdot \sin(x) \, dx = 0.5$$

Symbolically

$$\int_0^{\frac{\pi}{2}} \sin(x)^3 \, dx$$

$$\frac{2}{3}$$

Symbolically

$$\int_1^{e^2} \ln(x^2) \, dx$$

$$2 \cdot \exp(2) + 2 = 16.778$$

Symbolically

$$\int_0^{\frac{\pi}{2}} \left(1 - \frac{x^2}{2!} + \frac{x^4}{4!} - \frac{x^6}{6!} + \frac{x^8}{8!}\right) \cdot \sin(x) \, dx$$

$$\frac{1}{16} \cdot \pi^3 + \frac{1}{645120} \cdot \pi^7 - 2 \cdot \pi - \frac{1}{1920} \cdot \pi^5 + 5 = 0.5$$

Exhibit 3-24   **Exhibit 3-25**   **Exhibit 3-26**

Mathcad will change the result to feet for you. If you try to change feet to pounds, or some other conversion that does not make sense, Mathcad will not allow it. You can also use "Insert units" on the submenu of the "Units" menu, to assign units to a number. You should go ahead and experiment with the units in Mathcad. You will soon find that this powerful feature is easy to use, and a very handy tool to have at your disposal.

Numerically

$$\int_0^{\pi} \int_0^y \cos(x)^2 \, dx \, dy = 2.467$$

$$\int_0^{\infty} \frac{\cos(x)}{x^{\frac{1}{2}}} \, dx$$

$$\frac{d}{dx} \int_x^0 t^{-2 \cdot t} \, dt$$

Symbolically

$$\int_0^{\pi} \int_0^y \cos(x)^2 \, dx \, dy$$

$$\frac{1}{2} \cdot \sqrt{2} \cdot \sqrt{\pi}$$

$$-x^{(-2 \cdot x)}$$

$$\frac{1}{4} \cdot \pi^2 = 2.467$$

Exhibit 3-27   **Exhibit 3-28**   **Exhibit 3-29**

Numerically

$$\int_{1}^{3}\int_{1}^{z}\int_{1}^{y}\left(2\cdot\ln(x)+3+\frac{1}{x^3}\right)dx\,dy\,dz = 5.548$$

Symbolically

$$\int_{1}^{3}\int_{1}^{z}\int_{1}^{y}\left(2\cdot\ln(x)+3+\frac{1}{x^3}\right)dx\,dy\,dz$$

$$\frac{19}{2}\cdot\ln(3)-\frac{44}{9} = 5.548$$

Exhibit 3-30

Symbolically

$$\int_{\frac{1}{6}}^{\frac{1}{2}}\left|\frac{\cos(\pi\cdot x)^2}{\sin(\pi\cdot x)^4}\right|dx$$

$$\frac{\sqrt{3}}{\pi} = 0.551$$

Numerically

$$\int_{\frac{1}{6}}^{\frac{1}{2}}\left|\frac{\cos(\pi\cdot x)^2}{\sin(\pi\cdot x)^4}\right|dx = 0.551$$

Exhibit 3-31

Spiral problem 2: length of a compact-disk data spiral

Exhibit 3-41 is similar to the previous problem, except that here the length to be calculated is the length of the spiral of data on a compact disk. Here the thickness of the trail is much less than the thickness of tape in the previous example. In addition, we wanted the answer to be expressed in kilometers, so we retained the default SI system of units used by Mathcad.

Silo problem

Exhibit 3-42 is a problem involving a cow which has been tied to a silo and left to graze. As the diagram in the lower portion of the exhibit illustrates, the cow is free to graze away from the silo, within an arc with the radius of the length of the rope. However, as the cow approaches the silo, the physical existence of the silo itself becomes a constraint. The problem is to find the total area in which the cow can graze.

The problem involves integrating parametric equations. You can experiment with the example by changing

- The radius of the silo—the variable called "a" and set equal to 10 at the beginning of the exhibit
- The length of the rope—the variable called "b" and set equal to 20 in this exhibit

The length of the rope must be less than the silo radius times π to prevent overlap.

You will notice that some of the Mathcad expressions are included in the Mathcad document to do the calculations, while others are intended strictly to generate the diagram.

Using Complex Numbers

Imaginary numbers were invented to provide an answer to the question "What is the square root of a negative number?" Like much mathematics that was invented for purely theoretical reasons, imaginary (and complex) numbers have proved to be useful for many types of calculations in mathematics, science, and engineering.

As we have seen with other mathematical work, Mathcad provides an easy way to use the power of complex numbers.

In this section, we will first see how you can perform the standard arithmetic operations with complex numbers and later, how many common functions can also take complex numbers as arguments. We will also see some new functions, available in Mathcad, for working with complex numbers.

Symbolically

$$\int_0^\pi \sin(x)^4 \cdot \cos(x)^4 \, dx$$

$$\frac{3}{128} \cdot \pi = 0.074$$

Numerically

$$\int_0^\pi \sin(x)^4 \cdot \cos(x)^4 \, dx = 0.074$$

Exhibit 3-32

Symbolically

$$\int_0^{2\cdot\pi} \frac{1}{2} \cdot (1 + \cos(\theta))^2 \, d\theta$$

$$\frac{3}{2} \cdot \pi = 4.712$$

Numerically

$$\int_0^{2\cdot\pi} \frac{1}{2} \cdot (1 + \cos(\theta))^2 \, d\theta = 4.712$$

Exhibit 3-33 Area in polar coordinates by integration.

Symbolically

Numerically

$$\int_0^4 \int_{\frac{y^2}{4}}^{2\cdot\sqrt{y}} \int_0^{x+y^2} 1\, dz\, dx\, dy$$

$$A := \frac{2}{\sqrt{\pi}} \cdot \int_0^{.6} e^{-x^2} dx$$

$$\frac{1104}{35} = 31.543$$

$$A = 0.604$$

Symbolically

Numerically

$$A := \frac{2}{\sqrt{\pi}} \cdot \int_0^{.6} e^{-x^2} dx$$

$$\int_0^4 \int_{\frac{y^2}{4}}^{2\cdot\sqrt{y}} \int_0^{x+y^2} 1\, dz\, dx\, dy = 31.543$$

$$A := \frac{1.0703070536161574344}{\sqrt{\pi}}$$

$$A = 0.604$$

Exhibit 3-34

Exhibit 3-35

Entering complex numbers into Mathcad

To enter a complex number, type

- The real part of the number
- Followed by a plus sign "+"
- The imaginary part of the number
- The symbol for multiplication "*"

Then enter the symbol for the square root of -1 (usually denoted by i). You can select "i" from the arithmetic palette. See the examples in Exhibit 3-43. Alternatively, you can type the letter "i" from the keyboard immediately after the number. Once you have tried entering a few complex numbers and get acquainted with how it works, you may find this method to be simpler. If you use this method, be careful to omit the symbol "*" for multiplication, otherwise Mathcad will think you intend "i" to be a variable name. If you use this method, the only time you will need the "*" (or the palette symbol for "i") is to enter "i" itself, namely, "1i." Incidentally, if you prefer to use "j" to represent the square root of -1, you should go to the Format menu and select "Number." Near the bottom of this dialog box, you will be able to click on either "i" or "j."

Symbolically

$$V := \pi \cdot \int_{-2}^{2} \left(4 - y^2\right) dy$$

$$V := \frac{32}{3} \cdot \pi$$

$$\frac{32}{3} \cdot \pi = 33.51$$

Numerically

$$V := \pi \cdot \int_{-2}^{2} \left(4 - y^2\right) dy$$

$$V = 33.51$$

Exhibit 3-36

Compute the line integral

$$2 \cdot x \cdot y \cdot dx + \left(x^2 + y^2\right) \cdot dy$$

where the path for the integral is the parabola

$$y^2 = x$$

between (0,0) and (1,-1)

The integral becomes

$$\int_{0}^{1} \left(-4 \cdot t^4 - t^4 - t^2\right) dt$$

$$\frac{-4}{3}$$

Exhibit 3-37

You can also (see Exhibit 3-44) cause Mathcad to produce complex numbers as answers to some types of mathematical operations on real numbers—such as calculating the square root of a negative number.

Mathcad understands that the normal operations of arithmetic can take complex numbers as arguments and performs those operations without any special preparation from you, except to define the complex numbers that will be added, subtracted, multiplied, or divided. (See examples in Exhibit 3-45.)

Mathcad also recognizes that many other mathematical functions (including square roots, exponentials, logarithms, and trigonometric functions) can also take complex numbers as arguments and produce complex numbers as results (Exhibit 3-46). You can also use complex numbers in vectors and matrices. You will learn more about that in a later chapter.

There are also several Mathcad functions used specifically for complex numbers. These are the real, imaginary, arg, magnitude, and complex conjugate.

The real part of a complex number: Re(z)

In Exhibit 3-47, you see an example of how to use the Re(z) function to obtain the real part of a complex number. You might apply this in many physics problems where complex numbers are used to facilitate computation, but where the final result must be the real part of the

The rate of emission of radiation per unit wavelength interval
by each square meter of the surface of the sun is

$$R(\lambda) := \frac{3.74 \cdot 10^{-16}}{\lambda^5 \cdot \left(e^{\frac{2.52 \cdot 10^{-6}}{\lambda}} - 1 \right)} \quad \text{Watts/m}$$

Where the wavelength λ is in meters
We integrate over the range of visible wavelengths
To get the power in visible light

$$\int_{3.5 \cdot 10^{-7}}^{7 \cdot 10^{-7}} R(\lambda)\, d\lambda = 2.492 \cdot 10^7 \quad \text{Watts}$$

Exhibit 3-38

---

complex number. This function is easy to use: just type "Re(z)", and type the name of the complex number inside the parentheses.

The imaginary part of a complex number: Im(z)

This function gives the imaginary part of a complex number, just as Re(z) gives the real part. It is used in a similar way to the Re(z) function described above. An example is given in Exhibit 3-48.

The argument function: arg(z)

This function (see Exhibit 3-49) gives the angle (in radians) between the real axis and the complex number z. The angle will be between $-\pi$ and π.

The magnitude or absolute value of a complex number

To compute the magnitude of the complex number (see Exhibit 3-50), select the variable name in your Mathcad document (i.e., put a selection box around it) and use the "Vertical Line" key (above the backward slash)—it looks like a broken vertical line and appears on the key above the Enter key on many standard keyboards. You can use the absolute value symbol on the arithmetic palette as an alternative way of invoking this command.

Normal Distribution

$$F = \frac{1}{\left(\sigma \cdot \sqrt{2 \cdot \pi}\right)} \cdot \int_a^b e^{\left[-\frac{1}{2} \cdot \left(\frac{x - \mu}{\sigma}\right)^2\right]} dx$$

$\mu := 0$

$\sigma := 1$

$$F := \frac{1}{\left(\sigma \cdot \sqrt{2 \cdot \pi}\right)} \cdot \int_0^{2 \cdot \sigma} e^{\left[-\frac{1}{2} \cdot \left(\frac{x - \mu}{\sigma}\right)^2\right]} dx$$

$$F := \frac{1}{(2 \cdot \sigma)} \cdot \frac{\sqrt{2}}{\sqrt{\pi}} \cdot \left[\frac{-1}{2} \cdot \sigma \cdot \sqrt{2} \cdot \sqrt{\pi} \cdot \mathrm{erf}\left[\frac{1}{2} \cdot \sqrt{2} \cdot \frac{(-2 \cdot \sigma + \mu)}{\sigma}\right] + \frac{1}{2} \cdot \sigma \cdot \sqrt{2} \cdot \sqrt{\pi} \cdot \mathrm{erf}\left[\frac{1}{(2 \cdot \sigma)} \cdot \mu \cdot \sqrt{2}\right]\right]$$

$F = 0.477$

$2 \cdot .477 = 0.954$ For both tails of the normal distribution

Solving numerically
For three standard deviations

$$\frac{1}{\left(\sigma \cdot \sqrt{2 \cdot \pi}\right)} \cdot \int_0^{3 \cdot \sigma} e^{\left[-\frac{1}{2} \cdot \left(\frac{x - \mu}{\sigma}\right)^2\right]} dx = 0.499$$

$0.499 \cdot 2 = 0.998$ For both tails of the normal distribution

Exhibit 3-39

$t := .001 \cdot in$

$r1 := 0.5 \cdot in$

$r2 := 1 \cdot in$

Compute the length of a spiral of recording tape of thickness t, wound on a spool of radius r1. The outer radius of the spiral is r2.

You can compute the length of the tape by obtaining a formula for the differential of length in terms of an element of radius dr and integrating from the inner radius to the outer radius. You can see a more complete explanation of the method in Calculus for the Practical Worker by J. E. Thompson.

$$\int_{r1}^{r2} \sqrt{1 + \left(r^2\right) \cdot \left[\left[2 \cdot \left(\left(\frac{\pi}{t}\right)\right)\right]\right]^2} \, dr = 59.847 \bullet m$$

You may have realized that you do not have to do an integration to compute this length, any more than you need to integrate to find the area of a circle. You could notice that an equivalent result is obtained by dividing the "area" of tape by t

$$\frac{\pi \cdot \left(r2^2 - r1^2\right)}{t} = 59.847 \bullet m$$

Exhibit 3-40

The complex conjugate of a complex number

Exhibit 3-51 shows how to compute the complex conjugate (a complex number in which the sign of the imaginary part of the original number has been reversed). To use this command, after selecting the variable name, type the left (opening) quotation mark ("). Multiplying the original number by its complex conjugate and taking the square root of the answer gives a real number which is the magnitude of the original complex number.

Differentiating complex exponentials

Complex numbers, specifically complex exponentials, have long had a special place in many types of mathematical problems involving electromagnetic radiation or differential equations because of the special property that e^x remains unchanged under differentiation and integra-

Find the length of the data spiral on a compact disk

$$d := \frac{2.54 \cdot 2.38}{100} \cdot m$$

$$d = 0.06 \cdot m$$

$$a := 1.6 \cdot 10^{-6} \cdot m$$

$$a = 1.6 \bullet 10^{-6} \cdot m$$

$$c := \frac{2.54 \cdot .75}{100} \cdot m$$

$$ab := \int_{c}^{d} \sqrt{1 + \left(r^2\right) \cdot \left[\left[2 \cdot \left(\left(\frac{\pi}{a}\right)\right)\right]^2\right]} \, dr$$

$$ab = 6463 \bullet m$$

Or in kilometers

$$ab = 6.463 \circ km$$

Exhibit 3-41

tion. The result is that many types of problems become much simpler when expressed in terms of complex exponentials. See Exhibit 3-52 for a demonstration of this property and for the further demonstration of how multiplication increases (adds) the arguments of the numbers being added.

Schrödinger's wave equation

In quantum mechanics, one of the great theoretical triumphs of the twentieth century, complex numbers are no longer just a computational aid, but an integral part of the theory. In Exhibit 3-53, we see a simple example (see W. Heitler's book *Wave Mechanics* [Oxford University Press, 1956]) of how a wave equation with wavelength lambda for a wave along the x axis is differentiated twice to derive the Schrödinger wave equation, which lies at the heart of quantum mechanics.

What is the exact area that a cow can graze if it is tied to the end of a rope which is attached to a silo and there are no other obstructions to the cow's grazing?

silo := 10   (the radius of the silo)   rope := 20   (the length of the rope)

maxlen := silo·π   (maxlen = 31.416   feet - the maximum length of the rope without an overlap.)

rope := if(rope > maxlen, maxlen, rope)   (Do not allow the rope = 20   to cause an overlap)

$f := \dfrac{rope}{silo}$ (Where $\dfrac{f}{deg}$ = 114.592   degrees is the wrap angle of the rope around the silo.)

$$involute := \int_0^{rope} \frac{\frac{1}{2}\cdot(rope - t)^2}{silo}\,dt$$ involute portion of the grazing area

$halfcircle := \dfrac{\pi}{2}\cdot rope^2$ (half circle area portion of the grazing area)

area := involute·2 + halfcircle   (area = 894.985   square feet - total grazing area)

t := 0, .01 .. 1   (generic parameter)

sx(t) := silo·cos(2·π·t)
sy(t) := silo·sin(2·π·t) + silo   (the silo)

hx(t) := rope·cos(π·(1 + t))   (the half circle)
hy(t) := rope·sin(π·(1 + t))

ix(t) := silo·sin(t·f) + rope·(1 − t)·cos(t·f)   (parameteric equations defining
                                                       the involute of a circle)
iy(t) := silo·(1 − cos(t·f)) + rope·(1 − t)·sin(t·f)

━━━ silo
─── half circle
━━━ involute
━━━ involute

Exhibit 3-42

z1 := 1 + 2·i

z2 := 3 + 4·i

Addition

z1 + z2 = 4 + 6i

To enter a complex number

Subtraction

z2 − z1 = 2 + 2i

Example 1

$$\sqrt{-4} = 2i$$

Example 1

z1 := 2 + 3·i

Example 1

Multiplication

z1·z2 = −5 + 10i

Division

Example 2

z2 := 5 − 2.34·i

Example 2

acos(−5) = 3.142 − 2.292i

$$\frac{z1}{z2} = 0.44 + 0.08i$$

Exhibit 3-43

Exhibit 3-44

Exhibit 3-45

Some functions become multivalued when taking complex arguments

When taking complex numbers as arguments, some functions that would normally return a single value become multivalued. Roots and logarithms are examples of functions that take multiple values with complex arguments. You probably recall that numbers have three cube roots, four 4th roots, and so on when complex numbers are used. Mathcad's functions that take complex arguments, however, return only the principal value, the one having the smallest angle from the x axis. The one exception to this rule is the nth-root operator (accessible from the arithmetic palette), which always returns a real root when one is available. So if you raise a number to the one-third power (equivalent to taking the cube root), Mathcad returns the principal value, but if you use the nth-root operator (found on the arithmetic palette), Mathcad gives you a real root if possible. See Exhibit 3-54 for an example of this.

Setting the complex tolerance

On the Format menu when you select "Number," you will find the "Complex Tolerance" setting which has a default value of 10, but which allows you to select any integer between 0 and 63. This setting tells Mathcad that when either the real or the imaginary part of a complex number exceeds the other part by a ratio of the power of 10 that you have selected, Mathcad should

z1 := 5 + 7·i

ln(z1) = 2.152 + 0.951i

z2 := 4 − i

e^{z2} = 29.5 − 45.943i

z3 := .5 + .7·i

sin(z3) = 0.602 + 0.666i

z4 := 3 + 5·i

log(2) = 0.301

$\dfrac{\log(z4)}{\log(2)}$ = 2.544 + 1.487i   Is the log of z4 to the base 2

Exhibit 3-46

z1 := 1 + 2·i

Re(z1) = 1

Exhibit 3-47 The real part of a complex number.

z1 := 1 + 2·i

Im(z1) = 2

Exhibit 3-48 The imaginary part of a complex number.

---

ignore the smaller value for display purposes and treat the number as only real or only imaginary. Mathcad then rounds off the displayed number, but retains the full value internally. See Exhibit 3-55 for an example of this.

Exhibit 3-56 gives some examples of calculations using complex numbers. After defining a vector "b" with complex elements (based on an exponential function), the exhibit shows the results of applying the "Re" and the "Im" operators to produce vectors that include only the real parts of the elements of the original vector, and only the imaginary parts of the original vector, respectively. Then we look at the arguments (angle in polar representation) of the vector's elements, and we see how the arguments increase but the magnitude stays the same.

---

z1 := 1 + 2·i

arg(z1) = 1.107

Exhibit 3-49 The argument function.

z1 := 1 + 2·i

| z1 | = 2.236

Exhibit 3-50 The magnitude or absolute value of a complex number.

z1 := 1 + 2·i

$\overline{z1}$ = 1 − 2i

Exhibit 3-51 The complex conjugate of a complex number.

First derivative

$a = e^x$

$\dfrac{d}{dx} e^x$

exp(x)

Second derivative

$\dfrac{d^2}{dx^2} e^x$

exp(x)

Third derivative

$\dfrac{d^3}{dx^3} e^x$

exp(x)

Now define a and b

$a := e^{0.5 \cdot i}$ $\qquad\qquad$ $b := e^{0.2 \cdot i}$

And multiply

$c := a \cdot b$

$c = 0.765 + 0.644i$

$|a| = 1$	$\arg(a) = 0.500$	Notice that the multiplication
$|b| = 1$	$\arg(b) = 0.200$	of "a" times "b" adds their arguments (one of the reasons complex
$|c| = 1$	$\arg(c) = 0.700$	exponentials simplify many types of problems)

Exhibit 3-52 Differentiating complex exponentials.

Starting with a wave equation ψ

$$\psi = A \cdot e^{2 \cdot \pi \cdot i \cdot \frac{x}{\lambda}}$$

Differentiating once with respect to x

$$\frac{d}{dx} A \cdot e^{2 \cdot \pi \cdot i \cdot \frac{x}{\lambda}}$$

$$2i \cdot A \cdot \frac{\pi}{\lambda} \cdot \exp\left(2i \cdot \pi \cdot \frac{x}{\lambda}\right)$$

Differentiating a second time with respect to x

Or, equivalently

$$\frac{d}{dx} 2i \cdot A \cdot \frac{\pi}{\lambda} \cdot \exp\left(2i \cdot \pi \cdot \frac{x}{\lambda}\right) \qquad \frac{d^2}{dx^2} A \cdot e^{2 \cdot \pi \cdot i \cdot \frac{x}{\lambda}}$$

$$-4 \cdot A \cdot \frac{\pi^2}{\lambda^2} \cdot \exp\left(2i \cdot \pi \cdot \frac{x}{\lambda}\right) \qquad -4 \cdot A \cdot \frac{\pi^2}{\lambda^2} \cdot \exp\left(2i \cdot \pi \cdot \frac{x}{\lambda}\right)$$

So by differentiating twice we found that the second derivative of ψ w.r.t x is

equal to $-\left(2 \cdot \frac{\pi}{\lambda}\right)^2$ times ψ

So the waves satisfy the Schrodinger wave equation

$$\frac{d^2}{dx^2} \psi + \left(\frac{2 \cdot \pi}{\lambda}\right)^2 \cdot \lambda \quad \text{Is equal to zero}$$

Heitler says this "characteristic 'wave equation' is the basis of the whole [quantum] theory."

Exhibit 3-53 Schrödinger's wave equation.

Default Complex Tolerance

$z := -2$

$z := 1 + 10^{24} \cdot i$

Here Mathcad rounds
the real part to 0

$z = 1 \cdot 10^{24} i$

$z^{\frac{1}{5}} = 0.929 + 0.675i$

Complex Tolerance set at 63

$\sqrt[5]{z} = -1.149$

$z = 1 + 1 \cdot 10^{24} i$

Exhibit 3-54

Exhibit 3-55

We also look at the complex conjugates of the elements of the vector, and see the result of multiplying the complex conjugates of each element times the original element.

Functions that can take complex arguments in Mathcad

There are 21 functions that can take complex arguments in Mathcad:

sin	tanh
cos	csch
tan	sech
csc	coth
sec	asinh
cot	acosh
asin	atanh
acos	exp
atan	ln
sinh	log
cosh	

User-defined complex functions

You can define your own complex-valued functions in Mathcad in the same way you would create any other functions of your own. As an example, we will create a function that approximates the "LnGamma" function for a certain range of complex values. This user-defined Mathcad function (see Exhibit 3-57) calculates an approximation of the ln of the gamma function. At the end of the exhibit, we compare the result to a calculation using Mathcad's built-in ln and gamma functions.

$a := 3 + \sqrt{-1}$

$a = 3 + i$

$\text{Re}(a) = 3$

$x := 0 .. 7$

$b_x := 2 \cdot e^{\sqrt{-1} \cdot x}$

$$b = \begin{bmatrix} 2 \\ 1.081 + 1.683i \\ -0.832 + 1.819i \\ -1.98 + 0.282i \\ -1.307 - 1.514i \\ 0.567 - 1.918i \\ 1.92 - 0.559i \\ 1.508 + 1.314i \end{bmatrix}$$

Real part   Imaginary part

$d := \text{Re}(b)$ $f := \text{Im}(b)$

$$d = \begin{bmatrix} 2 \\ 1.081 \\ -0.832 \\ -1.98 \\ -1.307 \\ 0.567 \\ 1.92 \\ 1.508 \end{bmatrix} \qquad f = \begin{bmatrix} 0 \\ 1.683 \\ 1.819 \\ 0.282 \\ -1.514 \\ -1.918 \\ -0.559 \\ 1.314 \end{bmatrix}$$

Exhibit 3-56

argument (or angle)   magnitude (or absolute value)

$$c_x := \arg\left(b_x\right)$$

$$d_x := \left| b_x \right|$$

$$c = \begin{bmatrix} 0 \\ 1 \\ 2 \\ 3 \\ -2.283 \\ -1.283 \\ -0.283 \\ 0.717 \end{bmatrix}$$

$$d = \begin{bmatrix} 2 \\ 2 \\ 2 \\ 2 \\ 2 \\ 2 \\ 2 \\ 2 \end{bmatrix}$$

complex conjugate

$$e_x := \overline{b_x}$$

$$e = \begin{bmatrix} 2 \\ 1.081 - 1.683i \\ -0.832 - 1.819i \\ -1.98 - 0.282i \\ -1.307 + 1.514i \\ 0.567 + 1.918i \\ 1.92 + 0.559i \\ 1.508 - 1.314i \end{bmatrix}$$

complex conjugate times the original number

$$e_x \cdot b_x$$

$$\begin{bmatrix} 4 \\ 4 \\ 4 \\ 4 \\ 4 \\ 4 \\ 4 \\ 4 \end{bmatrix}$$

Exhibit 3-56 (*Continued*)

LG(x)=log(gamma(x))

approx both real and complex

for |z|<4 and Re(z)>0

$$a0 := \frac{1}{2} \cdot \ln(2 \cdot \pi)$$

$$a1 := \frac{1}{12}$$

$$a2 := \frac{1}{30}$$

$$a3 := \frac{53}{210}$$

$$a4 := \frac{195}{371}$$

$$a5 := 1.011523068$$

$$a6 := 1.517473649$$

$$LG(x) := \left(x - \frac{1}{2}\right) \cdot \ln(x) - x + a0 + \cfrac{a1}{x + \cfrac{a2}{x + \cfrac{a3}{x + \cfrac{a4}{x + \cfrac{a5}{x + a6}}}}}$$

$$LG(1 + 2 \cdot i) = -1.8761 + 0.1296i$$

Using Mathcad's built in functions, we get

$$\ln(\Gamma(1 + 2 \cdot i)) = -1.8761 + 0.1296i$$

Exhibit 3-57 User-defined complex functions.

Caution about plotting complex functions

While you can create a plot using complex numbers, you must be careful to understand what you want to plot and what Mathcad is going to do when you create your plot. Generally, you must specify the real or imaginary part of the complex numbers to be plotted. A more complete discussion of this is presented in Mathcad's Treasury, in the section on complex numbers.

Differentiation and Symbolic Calculations

In this chapter we see how to use Mathcad to do differentiation and symbolic calculations.

Numerical Derivatives

Mathcad can compute derivatives both numerically and symbolically. If you want to know the value of a function's derivative at a specific point (as you might to determine the slope at that point), you can use Mathcad's numerical derivative operator. To do this, you need to enter a value for the point at which the derivative is to be evaluated, then select the derivative operator from the "Calculus" palette. The derivative operator gives you placeholders to enter

1. The function to be differentiated
2. The variable with respect to which the derivative is to be computed

Once this information has been entered, click on the expression to select it and type the equal sign (=), which will cause Mathcad to return the answer. In the first part of Exhibit 4-1 you see an example of such a calculation.

The second example in Exhibit 4-1 is the same problem done by means of the symbolic processor. To do this symbolic calculation (which yields a mathematical expression instead of a number), we set up the problem the same way as in the preceding example and then select the expression to be evaluated and choose "Evaluate Symbolically" from the Symbolics menu.

An alternate way to do this symbolic derivative would have been to

- Type only the expression to be differentiated, "sin(x)" in this example, without the "d/dx."
- Then select the "x" in "sin(x)" to indicate that the derivative is to be computed with respect to "x."
- Then select "Variable" and "Differentiate" from the Symbolics menu.

For this example, the symbolic result of taking the derivative of "sin(x)" is "cos(x)," as you

Numerically

$$x := \frac{\pi}{4}$$

$$\frac{d}{dx} \sin(x) = 0.707$$

Symbolically

$$\frac{d}{dx} \sin(x)$$

$$\cos(x) = 0.707$$

Exhibit 4-1

---

would expect it to be. We then had Mathcad compute the numerical value to confirm that it checked with the numerical result computed in the first part of Exhibit 4-1.

Alignment of Segments of Text and Math

At this point, we should take note of what might be called a "housekeeping" matter. You may have noticed that Mathcad does not always align the parts of your work exactly as you might wish. For example, one line of text or equations will start a little to the left of another line, or part of the text and equations for one line of your Mathcad document will be slightly higher or lower than the other parts. To help you with this, Mathcad provides a second way of selecting your text and mathematics. The selection technique we have used previously in this book puts a solid line around an area of your work. There is another way to select equations and text. If you drag your mouse diagonally across several math and text regions of your work, Mathcad outlines the various text and mathematics regions with a dotted-line selection indicator. These selected regions could now be cut and pasted, just as you could do with the solid-line selection. Unlike the solid-line selection, however, the dotted-line selection indicator can encompass a number of math and text regions. This allows you to drag the selected parts of your Mathcad document around the screen as a group to improve the layout of your document. More important, it allows you to select areas that you wish to align and then use the "Align Regions" command on the "Format" menu to align the selected regions along the left margin ("Down") or in a horizontal line ("Across"), or to use the two tools on the toolbar which can do the same thing. These tools are near the middle of the toolbar and resemble bar graphs. One tool aligns selected regions along a horizontal line; the other creates a straight vertical alignment.

Exhibit 4-2 gives some additional examples of derivative calculations. You should note that

Example 1

$$x^3 - 3 \cdot x^2 + 2$$

by differentiation, yields

$$3 \cdot x^2 - 6 \cdot x$$

Notice here that the first and third examples are symbolic calculations. We have turned on the derivation comments so that Mathcad supplies a commentary about the procedures it is using to get the answers.

To do this, go to the "Symbolics" menu and select "Evaluation Style." Then click on "Show comments."

Example 2

$$y := .1$$

$$f(y) := 3 \cdot y^7 - 7 \cdot y^3 + 21 \cdot y^2$$

$$\frac{d}{dy} f(y) = 3.99002$$

Example 3

$$G(z) := 2 \cdot z + e^z \cdot i$$

by differentiation, yields

$$G(z) := 2 + i \cdot \exp(z)$$

Exhibit 4-2

Mathcad can also calculate derivatives of expressions involving complex numbers or user-defined functions.

If you expect to do many numerical derivative calculations, you will probably find it helpful to create a user-defined function to compute them. An example is shown in Exhibit 4-3. As you see in this example, you can use the derivative function you have defined to facilitate computation of first derivatives. Just specify the name of the function to be differentiated and the value at which the derivative is to be evaluated.

Higher-Order, Partial and Implicit Derivatives

Mathcad can also compute higher-order derivatives, partial derivatives, and derivatives of implicit functions. Higher-order derivatives can be calculated the same way as first-order

You can create your own function for computing a derivative

$$p(x) := x^5 - 2 \cdot x$$

$$x := 1.3$$

Define the function to take derivatives

$$G(x) := 2 \cdot x^7 + e^x \cdot i$$

$$D(f, x) := \frac{d}{dx} f(x)$$

$$\frac{d}{dx} G(x) = 67.575 + 3.669i$$

Then use it to compute derivatives of $p(x)$

$$\frac{d^2}{dx^2} G(x) = 311.886 + 3.669i$$

$$D(p, 2) = 78$$

$$D(p, .01) = -2$$

$$\frac{d^3}{dx^3} G(x) = 1.2 \cdot 10^3 + 3.669i$$

$$D\left(p, \frac{\pi}{10}\right) = -1.951$$

$$\frac{d^4}{dx^4} G(x) = 3.691 \cdot 10^3 + 3.669i$$

$$D(p, 10) = 5 \cdot 10^4$$

Exhibit 4-3

Exhibit 4-4

---

derivatives. This is done by repeating the same process a number of times. Mathcad provides a better way—instead of using the derivative operator "n" times to compute an order "n" derivative, Mathcad has an operator specifically created for this purpose. You can see an example of this in Exhibit 4-4. Symbolic higher-order derivatives are calculated as in Exhibit 4-5, by repeated use of the symbolic differentiation process described above. Alternatively, you can use the higher-order derivative operator to have Mathcad return the symbolic evaluation as shown in Exhibit 4-6. Note that here we used the right-pointing arrow found on the "Symbolic Keyword" palette—or by using the control key together with the period key (Ctrl + ".") instead of evaluating symbolically from the "Symbolics" menu.

Mathcad's derivative operators can also be used to compute partial derivatives as in Exhibit 4-7. Partial derivatives are used for functions that have more than one independent variable. The derivative is computed with respect to one variable while the other variables are treated as constants. Mathcad "understands" when you try to take a derivative with respect to a certain variable in a multivariate function, that the partial derivative is what is required and that it should be computed by holding the other variables constant. Notice that here we use two types of notation for partial derivatives. One notation looks like a rounded version of the standard derivative notation. An example is $\partial/\partial x$, which is to be interpreted as "the partial derivative with respect to x." Mathcad does not use this notation although it is found in many

$$f(x) := \left[x^7 + 3 \cdot e^{\left(x^2\right)} \right] - \sin(2 \cdot x)$$

First Derivative

$$\left[x^7 + 3 \cdot e^{\left(x^2\right)} \right] - \sin(2 \cdot x)$$

$$7 \cdot x^6 + 6 \cdot x \cdot \exp\left(x^2\right) - 2 \cdot \cos(2 \cdot x)$$

Second Derivative

$$7 \cdot x^6 + 6 \cdot x \cdot \exp\left(x^2\right) - 2 \cdot \cos(2 \cdot x)$$

$$42 \cdot x^5 + 6 \cdot \exp\left(x^2\right) + 12 \cdot x^2 \cdot \exp\left(x^2\right) + 4 \cdot \sin(2 \cdot x)$$

Third Derivative

$$42 \cdot x^5 + 6 \cdot \exp\left(x^2\right) + 12 \cdot x^2 \cdot \exp\left(x^2\right) + 4 \cdot \sin(2 \cdot x)$$

$$210 \cdot x^4 + 36 \cdot x \cdot \exp\left(x^2\right) + 24 \cdot x^3 \cdot \exp\left(x^2\right) + 8 \cdot \cos(2 \cdot x)$$

Fourth Derivative

$$210 \cdot x^4 + 36 \cdot x \cdot \exp\left(x^2\right) + 24 \cdot x^3 \cdot \exp\left(x^2\right) + 8 \cdot \cos(2 \cdot x)$$

$$840 \cdot x^3 + 36 \cdot \exp\left(x^2\right) + 144 \cdot x^2 \cdot \exp\left(x^2\right) + 48 \cdot x^4 \cdot \exp\left(x^2\right) - 16 \cdot \sin(2 \cdot x)$$

Exhibit 4-5

$$f(x) := \left[x^7 + 3 \cdot e^{\left(x^2\right)} \right] - \sin(2 \cdot x)$$

$$\frac{d^4}{dx^4} f(x) \rightarrow 840 \cdot x^3 + 36 \cdot \exp\left(x^2\right) + 144 \cdot x^2 \cdot \exp\left(x^2\right) + 48 \cdot x^4 \cdot \exp\left(x^2\right) - 16 \cdot \sin(2 \cdot x)$$

Exhibit 4-6

Partial Derivatives

$$x^3 + 7 \cdot x^2 \cdot y + 3 \cdot x \cdot y^5 - 5 \cdot y^6$$

f_x $\qquad\qquad\qquad\qquad\qquad$ f_y

$x^3 + 7 \cdot x^2 \cdot y + 3 \cdot x \cdot y^5 - 5 \cdot y^6$ $\qquad$ $x^3 + 7 \cdot x^2 \cdot y + 3 \cdot x \cdot y^5 - 5 \cdot y^6$

$3 \cdot x^2 + 14 \cdot x \cdot y + 3 \cdot y^5$ $\qquad\qquad$ $7 \cdot x^2 + 15 \cdot x \cdot y^4 - 30 \cdot y^5$

f_{xx} $\qquad\qquad\qquad\qquad\qquad$ f_{yy}

$3 \cdot x^2 + 14 \cdot x \cdot y + 3 \cdot y^5$ $\qquad\qquad$ $7 \cdot x^2 + 15 \cdot x \cdot y^4 - 30 \cdot y^5$

$6 \cdot x + 14 \cdot y$ $\qquad\qquad\qquad\quad$ $60 \cdot x \cdot y^3 - 150 \cdot y^4$

f_{xxx} $\qquad\qquad\qquad\qquad\qquad$ f_{yyy}

$6 \cdot x + 14 \cdot y$ $\qquad\qquad\qquad\quad$ $60 \cdot x \cdot y^3 - 150 \cdot y^4$

6 $\qquad\qquad\qquad\qquad\qquad\quad$ $180 \cdot x \cdot y^2 - 600 \cdot y^3$

$f_{xy} = f_{yx}$

$3 \cdot x^2 + 14 \cdot x \cdot y + 3 \cdot y^5$

$14 \cdot x + 15 \cdot y^4$

$f_{xxy} = f_{xyx} = f_{yxx}$

$6 \cdot x + 14 \cdot y$

14

$f_{xyy} = f_{yxy} = f_{yyx}$

$14 \cdot x + 15 \cdot y^4$

$60 \cdot y^3$

Exhibit 4-7

textbooks. The alternative notation for the same derivative would be F_x (or the second partial derivative with respect to x would be designated as F_{xx}).

The directional derivative gives the change of a multivariate function in a given direction. Computation of the directional derivative is illustrated in Exhibit 4-8.

Creation of a user-defined function to compute the curvature of a function at a given point is illustrated in Exhibit 4-9. As with the function we defined above to compute the derivative, this allows the computation of the curvature by simply inputting the name of the function and the point at which it is to be evaluated.

An example of the chain rule for partial derivatives is given in Exhibit 4-10.

Another partial derivative problem involving the chain rule is illustrated in Exhibit 4-11.

This problem involves a function that is not explicitly defined in terms of a dependent variable. So we need to know how to compute dy/dx for an implicitly defined function. This is shown in Exhibit 4-12.

Here we show how to compute the tangent plane and the normal line to a surface using Mathcad's ability to compute partial derivatives. This is illustrated in Exhibit 4-13.

More partial derivative calculations are shown in Exhibit 4-14.

Many problems involving fields require the use of one or more of the following vector operators:

- The gradient is illustrated in Exhibit 4-15. For a scalar function $u(x,y,z)$ the gradient can be defined geometrically as a vector whose direction is the direction in which u increases most rapidly and whose magnitude is the *directional derivative* of u, that is, the rate of increase of u per unit distance in that direction.

- The divergence is illustrated in Exhibit 4-16. The divergence can be thought of as relating the amount of some substance (fluid, for example) inside an enclosed volume to the flux of that substance across the surface of the volume.

- The curl is illustrated in Exhibit 4-17. The curl of a vector function $A(x,y,z)$ at any point is a measure of the extent to which the vector function A circles around that point. An example is the magnetic field around a wire carrying an electric current, where the curl of the magnetic field intensity is proportional to the current density.

Here is another example of how Mathcad can help you with problems involving implicit differentiation. See Exhibit 4-18.

Symbolic Results

The remainder of this chapter deals with the use of the symbolic processor. Mathcad (especially the Professional version) gives you a powerful array of operations that can be done symbolically. There are three ways of performing symbolic operations in Mathcad:

- Use the symbolic equal sign. This feels much like doing numerical math. Note that the symbolic equal sign is a "live" operator. See "Live Symbolics" later in this chapter.

- Use commands from the Symbolics menu or use symbolic keywords together with the symbolic equal sign. This gives you greater control over the symbolic transformation.

Compute the Directional Derivative of

$$2 \cdot x^3 \cdot y - 3 \cdot y^2 \cdot z \qquad \text{at (1,2,-1) in the direction toward (3,-1,5)}$$

Compute f_x

$$2 \cdot x^3 \cdot y - 3 \cdot y^2 \cdot z$$

$$6 \cdot x^2 \cdot y$$

Compute f_y

$$2 \cdot x^3 \cdot y - 3 \cdot y^2 \cdot z$$

$$2 \cdot x^3 - 6 \cdot y \cdot z$$

Compute f_z

$$2 \cdot x^3 \cdot y - 3 \cdot y^2 \cdot z$$

$$-3 \cdot y^2$$

Evaluate at (1,2,-1)

$$x := 1 \qquad y := 2 \qquad z := -1$$

$$6 \cdot x^2 \cdot y = 12 \qquad 2 \cdot x^3 - 6 \cdot y \cdot z = 14 \qquad -3 \cdot y^2 = -12$$

So $a = 6 \cdot x^2 \cdot y \cdot i + \left(2 \cdot x^3 - 6 \cdot y \cdot z \right) \cdot j + \left(-3 \cdot y^2 \right) \cdot k$ \qquad Or 12*i+14*j-12*k

$$a := \begin{bmatrix} 12 \\ 14 \\ -12 \end{bmatrix}$$

Exhibit 4-8

And the unit vector from (1,2,-1) to (3,-1,5) is

$$\frac{(3 - 1)\cdot i + (-1 - 2)\cdot j + (5 - (-1))\cdot k}{\sqrt{2^2 + (-3)^2 + 6^2}}$$

$$\frac{2}{7}\cdot i - \frac{3}{7}\cdot j + \frac{6}{7}\cdot k$$

$$b := \begin{bmatrix} \dfrac{2}{7} \\[2mm] -\dfrac{3}{7} \\[2mm] \dfrac{6}{7} \end{bmatrix}$$

Now the directional derivative is computed by multiplying (dot product) the vector a computed above times the unit vector b

$a\cdot b = -12.857$

Exhibit 4-8 (*Continued*)

- Use "Symbolic optimization." Refer to the User's Guide if you want to see a description of this process for getting more efficient solutions by making the numerical and symbolic processors work together.

The Symbolics Menu

To make it easier to use the "Symbolics" menu, the operations on the menu are grouped so that related ones are together, and some also have submenus. Remember that commands from the "Symbolics" menu are not "live." Here we list the operations on the Symbolics menu. Then a brief explanation of each operation will follow.

Symbolics Menu:
   Symbolic Operators that work on an expression
      Evaluate
         Symbolically

Compute a user defined function that will compute the curvature of a function at a point x.

$$C(f,x) := \frac{\dfrac{d^2}{dx^2}f(x)}{\left[1 + \left(\dfrac{d}{dx}f(x)\right)^2\right]^{\frac{3}{2}}}$$

$F(x) := 5 \cdot x + .3$

$C(F,2) = 0$ The curvature is 0 since this is a straight line.

Or another example

$Z(x) := 2 \cdot x^2 + 5 \cdot x + .3$

$C(Z,2) = 1.805 \cdot 10^{-3}$

Exhibit 4-9

Floating Point
Complex
Simplify
Expand
Factor
Collect
Polynomial Coefficients
Symbolic Operators that work on a variable
Variable
Solve
Substitute
Differentiate
Integrate
Expand to Series
Convert to Partial Fraction
Symbolic Operators that work on a matrix
Matrix

Given w=x·y + z and x=cos(t), y=sin(t), z=t

Compute $\dfrac{dw}{dt}$

w_x	x·y + z	dx/dt	cos(t)
	y		-sin(t)
w_y	x·y + z	dy/dt	sin(t)
	x		cos(t)
w_z	x·y + z	dz/dt	t
	1		1

Then $\dfrac{dw}{dt} = \dfrac{dx}{dt} + \dfrac{dy}{dt} + \dfrac{dz}{dt}$

y·-sin(t) + x·cos(t) + 1·1

-y·sin(t) + x·cos(t) + 1

Exhibit 4-10 Chain rule for partial derivatives.

    Transpose
    Invert
    Determinant
Symbolic Operators that perform transforms
    Transforms
        Fourier Transform
        Inverse Fourier Transform
        Laplace Transform
        Inverse Laplace Transform
        Z Transform
        Inverse Z Transform
Evaluation Style

Given that $z = x^2 \cdot y + e^{x \cdot y}$, $x = t^2$ and $y = \ln(t)$.

Compute $\dfrac{dz}{dt} = \dfrac{\partial z}{\partial x} \cdot \dfrac{dx}{dt} + \dfrac{\partial z}{\partial y} \cdot \dfrac{dy}{dt}$

$\dfrac{\partial z}{\partial x}$ is $x^2 \cdot y + e^{x \cdot y}$

$2 \cdot x \cdot y + y \cdot \exp(x \cdot y)$

$\dfrac{dx}{dt}$ is t^2

$2 \cdot t$

$\dfrac{\partial z}{\partial y}$ is $x^2 \cdot y + e^{x \cdot y}$

$x^2 + x \cdot \exp(x \cdot y)$

$\dfrac{dy}{dt}$ is $\ln(t)$

$\dfrac{1}{t}$

So $\dfrac{dz}{dt}$ is $(2 \cdot x \cdot y + y \cdot \exp(x \cdot y)) \cdot (2 \cdot t) + \left(x^2 + x \cdot \exp(x \cdot y)\right) \cdot \dfrac{1}{t}$

Exhibit 4-11 Chain rule for partial derivatives.

Given $x^2 + y^2 \cdot \tan(x) + e^y = 0$

find $\dfrac{dy}{dx}$

$\dfrac{dy}{dx} = -\dfrac{f_x}{f_y}$

f_x is $x^2 + y^2 \cdot \tan(x) + e^y$

$2 \cdot x + y^2 \cdot \left(1 + \tan(x)^2\right)$

f_y is $x^2 + y^2 \cdot \tan(x) + e^y$

$2 \cdot y \cdot \tan(x) + \exp(y)$

So $\dfrac{dy}{dx}$ is $\dfrac{-\left[2 \cdot x + y^2 \cdot \left(1 + \tan(x)^2\right)\right]}{2 \cdot y \cdot \tan(x) + \exp(y)}$

Exhibit 4-12

The steps for using the commands on the Symbolics menu are:

- Place the mathematical expression between two editing lines (or for some commands, click on a variable in the expression)
- Choose the command you want to use from the Symbolics menu

Symbolic Operators that Work on an Expression

When you select a member of this group of operations the expression is treated as a unit by the symbolic processor. So proceed as explained above, by placing the expression between two

Given $F(x, y, z) = \left(\dfrac{x^2}{a^2} + \dfrac{y^2}{b^2} + \dfrac{z^2}{c^2} \right) - 1$

and given that a=1.59   b=2.23   c=1.41
compute the equation of the tangent plane and the
normal line to the above ellipsoid at (1.1, 1.3, .6)

x1 := 1.1   y1 := 1.3   z1 := .6

a := 1.59   b := 2.23   c := 1.41

F_x $\left(\dfrac{x^2}{a^2} + \dfrac{y^2}{b^2} + \dfrac{z^2}{c^2} \right) - 1$

$2 \cdot \dfrac{x}{a^2}$ At x equals 1.1 this is $\dfrac{2 \cdot x1}{a^2} = 0.87$

F_y $\left(\dfrac{x^2}{a^2} + \dfrac{y^2}{b^2} + \dfrac{z^2}{c^2} \right) - 1$

$2 \cdot \dfrac{y}{b^2}$ At y equals 1.3 this is $\dfrac{2 \cdot y1}{b^2} = 0.523$

F_z $\left(\dfrac{x^2}{a^2} + \dfrac{y^2}{b^2} + \dfrac{z^2}{c^2} \right) - 1$

$2 \cdot \dfrac{z}{c^2}$ At z equals .6 this is $\dfrac{2 \cdot z1}{c^2} = 0.604$

Then the tangent plane is $2 \cdot \dfrac{x1}{a^2} \cdot (x - x1) + 2 \cdot \dfrac{y1}{b^2} \cdot (y - y1) + 2 \cdot \dfrac{z}{c^2} \cdot (z - z1) = 0$

$0.87 \cdot (x - 1.1) + 0.523 \cdot (y - 1.3) + 0.604 \cdot (z - .6)$

Or

$.87 \cdot x - 1.9993 + .523 \cdot y + .604 \cdot z = 0$

Exhibit 4-13

The normal line is $\dfrac{x - x1}{\left(\dfrac{x1}{a^2}\right)} = \dfrac{y - y1}{\left(\dfrac{y1}{b^2}\right)} = \dfrac{z - z1}{\left(\dfrac{z1}{c^2}\right)}$

$\dfrac{x1}{a^2} = 0.435$ $\dfrac{y1}{b^2} = 0.261$ $\dfrac{z1}{c^2} = 0.302$

Or $\dfrac{x - 1.1}{.435} = \dfrac{y - 1.3}{.261} = \dfrac{z - .6}{.302}$

Exhibit 4-13 (*Continued*)

Compute f_x

$\sqrt{x^2 + y^2}$

$\dfrac{1}{\sqrt{x^2 + y^2}} \cdot x$

Compute f_y

$\sqrt{x^2 + y^2}$

$\dfrac{1}{\sqrt{x^2 + y^2}} \cdot y$

Compute f_{xx}

$\dfrac{1}{\sqrt{x^2 + y^2}} \cdot x$

$\dfrac{-1}{\left(x^2 + y^2\right)^{\left(\frac{3}{2}\right)}} \cdot x^2 + \dfrac{1}{\sqrt{x^2 + y^2}}$

Compute f_{yy}

$\dfrac{1}{\sqrt{x^2 + y^2}} \cdot y$

$\dfrac{-1}{\left(x^2 + y^2\right)^{\left(\frac{3}{2}\right)}} \cdot y^2 + \dfrac{1}{\sqrt{x^2 + y^2}}$

Exhibit 4-14

If $u = x^2 \cdot y \cdot z^3$

Compute the gradient of u

The gradient is equal to $i \cdot \dfrac{\partial}{\partial x} + j \cdot \dfrac{\partial}{\partial y} + k \cdot \dfrac{\partial}{\partial z}$

$\dfrac{\partial}{\partial x}$ is equal to $\qquad x^2 \cdot y \cdot z^3$

$\qquad\qquad\qquad\qquad 2 \cdot x \cdot y \cdot z^3$

$\dfrac{\partial}{\partial y}$ is equal to $\qquad x^2 \cdot y \cdot z^3$

$\qquad\qquad\qquad\qquad x^2 \cdot z^3$

$\dfrac{\partial}{\partial z}$ is equal to $\qquad x^2 \cdot y \cdot z^3$

$\qquad\qquad\qquad\qquad 3 \cdot x^2 \cdot y \cdot z^2$

So the gradient is equal to $\qquad 2 \cdot x \cdot y \cdot z^3 \cdot i + x^2 \cdot z^3 \cdot j + 3 \cdot x^2 \cdot y \cdot z^2 \cdot k$

Exhibit 4-15

---

editing lines and then choosing the symbolic operation you want Mathcad to perform from the Symbolics menu.

Evaluate

The important thing to understand about the symbolic processor is that it computes exact answers.

Symbolically

Because it works with exact results, the symbolic processor treats fractions as fractions and not as decimal approximations. An example of this is seen in Exhibit 4-19.

For vector $A = 3 \cdot x^2 \cdot i + 5 \cdot x \cdot y^2 \cdot j + x \cdot y \cdot z^3 \cdot k$ at Point (1,2,3)

Compute the divergence of A

$$\text{div } A = \frac{\partial A_x}{\partial x} + \frac{\partial A_y}{\partial y} + \frac{\partial A_z}{\partial z}$$

$3 \cdot x^2$

$6 \cdot x$

$5 \cdot x \cdot y^2$

$10 \cdot x \cdot y$

$x \cdot y \cdot z^3$

$3 \cdot x \cdot y \cdot z^2$

Divergence $\qquad 6 \cdot x + 10 \cdot x \cdot y + 3 \cdot x \cdot y \cdot z^2$

So at (1,2,3)

$x := 1 \qquad y := 2 \qquad z := 3$

$6 \cdot x + 10 \cdot x \cdot y + 3 \cdot x \cdot y \cdot z^2 = 80$

Exhibit 4-16

Floating point

Although it is primarily designed to give you "exact" answers, the symbolic processor also "knows" how to produce highly accurate floating point numerical results (up to 4000 places of accuracy). When you choose "Floating Point" from the submenu of "Evaluation" on the Symbolics menu, Mathcad will ask you to insert the number of places to which the evaluation should be carried. These calculations are also selected from the "Symbolics" menu. In Exhibit 4-20, we see three

Compute the curl of $A = i \cdot (x \cdot y \cdot z) + j \cdot \left(x \cdot y \cdot z^2\right) + k \cdot \left(x^3 \cdot y \cdot z\right)$

Curl can be represented as the determinant of

$$\begin{bmatrix} i & j & k \\ \dfrac{\partial}{\partial x} & \dfrac{\partial}{\partial y} & \dfrac{\partial}{\partial z} \\ A_x & A_y & A_z \end{bmatrix}$$

Or $\quad i \cdot \left(\dfrac{\partial}{\partial y} \cdot A_z - \dfrac{\partial}{\partial z} \cdot A_y \right) + j \cdot \left(\dfrac{\partial}{\partial z} \cdot A_x - \dfrac{\partial}{\partial x} \cdot A_z \right) + k \cdot \left(\dfrac{\partial}{\partial x} \cdot A_y - \dfrac{\partial}{\partial y} \cdot A_x \right)$

$\dfrac{\partial}{\partial y} \cdot A_z$ is equal to $\qquad \left(x^3 \cdot y \cdot z\right)$

$\qquad\qquad\qquad\qquad\qquad x^3 \cdot z$

$\dfrac{\partial}{\partial z} \cdot A_y$ is equal to $\qquad x \cdot y \cdot z^2$

$\qquad\qquad\qquad\qquad\qquad 2 \cdot x \cdot y \cdot z$

$\dfrac{\partial}{\partial z} \cdot A_x$ is equal to $\qquad (x \cdot y \cdot z)$

$\qquad\qquad\qquad\qquad\qquad x \cdot y$

$\dfrac{\partial}{\partial x} \cdot A_z$ is equal to $\qquad \left(x^3 \cdot y \cdot z\right)$

$\qquad\qquad\qquad\qquad\qquad 3 \cdot x^2 \cdot y \cdot z$

$\dfrac{\partial}{\partial x} \cdot A_y$ is equal to $\qquad \left(x \cdot y \cdot z^2\right)$

$\qquad\qquad\qquad\qquad\qquad y \cdot z^2$

$\dfrac{\partial}{\partial y} \cdot A_x$ is equal to $\qquad (x \cdot y \cdot z)$

$\qquad\qquad\qquad\qquad\qquad y \cdot z$

So the curl is

$i \cdot \left(x^3 \cdot z - 2 \cdot x \cdot y \cdot z\right) + j \cdot \left(x \cdot y - 3 \cdot x^2 \cdot y \cdot z\right) + k \cdot \left(y \cdot z^2 - y \cdot z\right)$

Exhibit 4-17

We want to use implicit differentiation to find $\dfrac{dy}{dx}$

where $\ln\left(\dfrac{y}{x}\right) + y \cdot e^x = 0$

We compute f_x

$$\ln\left(\dfrac{y}{x}\right) + y \cdot e^x$$

$$\dfrac{-1}{x} + y \cdot \exp(x)$$

And we compute f_y

$$\ln\left(\dfrac{y}{x}\right) + y \cdot e^x$$

$$\dfrac{1}{y} + \exp(x)$$

We now find $\dfrac{dy}{dx}$ by dividing f_x by f_y and changing the sign

$$\dfrac{-1 \cdot \left(\dfrac{-1}{x} + y \cdot \exp(x)\right)}{\dfrac{1}{y} + \exp(x)}$$

Or

$$\dfrac{\dfrac{1}{x} - y \cdot \exp(x)}{\dfrac{1}{y} + \exp(x)}$$

Which simplifies to

$$-(-1 + \exp(x) \cdot y \cdot x) \cdot \dfrac{y}{(x \cdot (1 + y \cdot \exp(x)))}$$

Exhibit 4-18

An investor buys 100 shares of the stock of XYZ Corporation at 33 and 15/16 Dollars. A few days later, the stock price falls to 27 and 1/4. The investor decides that the XYZ Corporation is still a good company to invest in, and at the new (lower) price is even more attractive — so she buys another 100 shares. Ignoring brokerage commissions, what is the average purchase price of the 200 shares?

$$\frac{\left(33 + \frac{15}{16}\right) + \left(27 + \frac{1}{4}\right)}{2}$$

Using the symbolic processor to "evaluate" this expression, we get the average price as a fraction

$$\frac{979}{32}$$

To convert this to $ and fractional $, we compute

$$\text{floor}\left(\frac{979}{32}\right) = 30$$

For the number of dollars

And

$$\text{mod}(979, 32) = 19$$

For the number of 32nds

So the average purchase price is $ 30 $\frac{19}{32}$

Exhibit 4-19

---

different results returned by the symbolic processor for the square root of 17. As you see in the exhibit, the way you pose the question will determine which of the three answers you get.

In Exhibit 4-21 we illustrate calculations using the symbolic processor's floating point capability (to 75 decimal places). Use "Evaluate" on the Symbolics menu and then choose "Floating Point" on the submenu.

Complex

If you select "Complex" from the submenu of "Evaluate" on the Symbolics menu, Mathcad will attempt to evaluate your expression and return results that have the real and imaginary parts separated in the way you would normally expect. For an example, see Exhibit 4-22.

Symbolic Evaluation

$$\sqrt{17}$$

$$\sqrt{17}$$

Symbolic Evaluation

$$\sqrt{17.0}$$

4.1231056256176605498

Floating Point Evaluation to 60 places

$$\sqrt{17.0}$$

4.12310562561766054982140985597407702514719922537362043439863

Exhibit 4-20

π

3.14159265358979323846264338327950288419716939937510582097494459230781640629

e

2.71828182845904523536028747135266249775724709369995957496696762772407663035

acos(.01234)

1.55845601359328364428276104054835681122942605448009957444271483765423638538

Exhibit 4-21

$e^{i \cdot 2}$

$\cos(2) + i \cdot \sin(2)$

Exhibit 4-22

$(1+x)^7$ expand $\rightarrow 1 + 7 \cdot x + 21 \cdot x^2 + 35 \cdot x^3 + 35 \cdot x^4 + 21 \cdot x^5 + 7 \cdot x^6 + x^7$

$\dfrac{2}{3} + \dfrac{1}{2} + \dfrac{3}{7}$ simplify $\rightarrow \dfrac{67}{42}$

$\sin(x)^2 + \cos(x)^2$ simplify $\rightarrow 1$

Exhibit 4-23

Simplify

Mathcad tries to use the rules of algebra and trigonometry to put the expression in a simpler form as seen in Exhibit 4-23.

Expand

If you enter an expression such as $(1+x)11$ Mathcad will multiply it out, just as you learned to do in algebra class. See Exhibit 4-24.

Factor

Another thing you learned when you studied algebra was how to separate a complicated expression into its factors, essentially the opposite of "Expand Expression." Mathcad can do this for you effortlessly. See Exhibit 4-25.

$(1+x)^{11}$

$1 + 11 \cdot x + 55 \cdot x^2 + 165 \cdot x^3 + 330 \cdot x^4 + 462 \cdot x^5 + 462 \cdot x^6 + 330 \cdot x^7 + 165 \cdot x^8 + 55 \cdot x^9 + 11 \cdot x^{10} + x^{11}$

Or

$(1+x)^{11}$ expand $\rightarrow 1 + 11 \cdot x + 55 \cdot x^2 + 165 \cdot x^3 + 330 \cdot x^4 + 462 \cdot x^5 + 462 \cdot x^6 + 330 \cdot x^7 + 165 \cdot x^8 + 55 \cdot x^9 + 11 \cdot x^{10} + x^{11}$

Exhibit 4-24

$$1 + 11 \cdot x + 55 \cdot x^2 + 165 \cdot x^3 + 330 \cdot x^4 + 462 \cdot x^5 + 462 \cdot x^6 + 330 \cdot x^7 + 165 \cdot x^8 + 55 \cdot x^9 + 11 \cdot x^{10} + x^{11}$$

$$(x + 1)^{11}$$

Or

$$1 + 11 \cdot x + 55 \cdot x^2 + 165 \cdot x^3 + 330 \cdot x^4 + 462 \cdot x^5 + 462 \cdot x^6 + 330 \cdot x^7 + 165 \cdot x^8 + 55 \cdot x^9 + 11 \cdot x^{10} + x^{11} \text{ factor } \rightarrow (1 + x)^1$$

Exhibit 4-25

Collect

This operation collects terms containing like powers of the subexpression you select. See Exhibit 4-26. In this example, we selected x and Mathcad collected like powers of x.

Polynomial coefficients

Many expressions can be rewritten as polynomials. See Exhibit 4-27. Mathcad Professional will determine the coefficients for the polynomial when you enter the expression in your Mathcad document and select the variable in which the polynomial is to be expressed.

Symbolic Operators that Work on a Variable

Variable

By contrast, when you use a member of the Variable Group of operations, you need to specify a variable on which the operation is to be done. So you enter the expression and then select

---

$$\sin(x) + 3 \cdot \sin(x)^2 + 7 \cdot \sin(x)^3 + 11 \cdot \sin(x)^5$$

select $\sin(x)$

$$\begin{bmatrix} 0 \\ 1 \\ 3 \\ 7 \\ 0 \\ 11 \end{bmatrix}$$

$$x^3 + 2 \cdot y \cdot x^3 + 5 \cdot y^5 \cdot x^2 - x + 24 \cdot x \cdot y^3$$

$$(2 \cdot y + 1) \cdot x^3 + 5 \cdot y^5 \cdot x^2 + (24 \cdot y^3 - 1) \cdot x$$

Exhibit 4-26   **Exhibit 4-27**

$$\left(x^3 - 2{\cdot}x\right)$$

$$\begin{bmatrix} 0 \\ \sqrt{2} \\ -\sqrt{2} \end{bmatrix}$$

Exhibit 4-28

the variable that is key to this operation (integration with respect to x, for example) by clicking on it in the expression.

Solve

Exhibits 4-28 and 4-29 give examples of how to solve for a variable. Mathcad also enables you to solve systems of equations symbolically. You should keep in mind that finding symbolic

$$\left(x^3 - 3{\cdot}x\right) + 5 = 0$$

$$\begin{bmatrix} -\left(\dfrac{5}{2} + \dfrac{1}{2}{\cdot}\sqrt{21}\right)^{\left(\frac{1}{3}\right)} - \dfrac{1}{\left(\dfrac{5}{2} + \dfrac{1}{2}{\cdot}\sqrt{21}\right)^{\left(\frac{1}{3}\right)}} \\[4em] \dfrac{1}{2}{\cdot}\left(\dfrac{5}{2} + \dfrac{1}{2}{\cdot}\sqrt{21}\right)^{\left(\frac{1}{3}\right)} + \dfrac{1}{\left[2{\cdot}\left(\dfrac{5}{2} + \dfrac{1}{2}{\cdot}\sqrt{21}\right)^{\left(\frac{1}{3}\right)}\right]} + \dfrac{1}{2}{\cdot}i\cdot\sqrt{3}{\cdot}\left[-\left(\dfrac{5}{2} + \dfrac{1}{2}{\cdot}\sqrt{21}\right)^{\left(\frac{1}{3}\right)} + \dfrac{1}{\left(\dfrac{5}{2} + \dfrac{1}{2}{\cdot}\sqrt{21}\right)^{\left(\frac{1}{3}\right)}}\right] \\[4em] \dfrac{1}{2}{\cdot}\left(\dfrac{5}{2} + \dfrac{1}{2}{\cdot}\sqrt{21}\right)^{\left(\frac{1}{3}\right)} + \dfrac{1}{\left[2{\cdot}\left(\dfrac{5}{2} + \dfrac{1}{2}{\cdot}\sqrt{21}\right)^{\left(\frac{1}{3}\right)}\right]} - \dfrac{1}{2}{\cdot}i\cdot\sqrt{3}{\cdot}\left[-\left(\dfrac{5}{2} + \dfrac{1}{2}{\cdot}\sqrt{21}\right)^{\left(\frac{1}{3}\right)} + \dfrac{1}{\left(\dfrac{5}{2} + \dfrac{1}{2}{\cdot}\sqrt{21}\right)^{\left(\frac{1}{3}\right)}}\right] \end{bmatrix} = \begin{bmatrix} -2.279 \\ 1.14 - 0.946i \\ 1.14 + 0.946i \end{bmatrix}$$

Exhibit 4-29

Substitute $x + 1$ for z

$$z^3 + 4 \cdot z^2 + 5 \cdot z + 7$$

$$(x + 1)^3 + 4 \cdot (x + 1)^2 + 5 \cdot x + 12$$

Then Simplfy

$$x^3 + 7 \cdot x^2 + 16 \cdot x + 17$$

Exhibit 4-30

---

solutions to equations and systems of equations is a difficult process. It is possible for you to construct problems that cannot be solved by the symbolic processor.

Substitute

This is used to substitute an expression such as $(x+1)$ for the variable. See Exhibit 4-30.

Differentiate

Differentiation was covered extensively in this chapter. Here is a review of the differentiation procedure:

- enter the expression to be differentiated
- select an occurrence of the variable with respect to which the derivative is to be computed
- click on "Variable" and then on "Differentiate" and Mathcad will return the derivative

Integrate

Integration is covered in Chapter 3. Here is a review of the symbolic integration procedure:

- enter the expression to be integrated
- select an occurrence of the variable with respect to which the integral is to be computed
- click on "Variable" and then on "Integrate" and Mathcad will return the integral

e^x

$$1 + x + \frac{1}{2} \cdot x^2 + \frac{1}{6} \cdot x^3 + \frac{1}{24} \cdot x^4 + \frac{1}{120} \cdot x^5 + \frac{1}{720} \cdot x^6 + \frac{1}{5040} \cdot x^7 + \frac{1}{40320} \cdot x^8 + \frac{1}{362880} \cdot x^9 + O\left(x^{10}\right)$$

Exhibit 4-31

Expand to series

Select a variable. Mathcad will ask for the order of the series to be created. Mathcad then expands the given expression into a series. See Exhibit 4-31.

Convert to partial fraction

Select a variable in the denominator of the expression. Mathcad will try to factor the denominator into linear or quadratic factors having integer coefficients and then expand the expression into a sum of fractions with these factors as denominators. Partial fraction expansions are often used in the solution of difficult integration problems. See Exhibit 4-32.

Symbolic Operators that Work on a Matrix

Matrix

The Matrix Group allows you to do symbolic operations on matrices. The symbolic form of transpose a matrix, invert a matrix and compute the determinant of a matrix are selected from the "Matrix" section of the Symbolics menu. Matrix operations are discussed in detail in Chapter 6.

Transforms

The Transforms Group puts the power of Mathcad at your disposal to do Fourier, Laplace and Z Transforms and their inverses. We will illustrate only the Laplace Transform here (see

---

$$\frac{x^2 - 5 \cdot x - 3}{x^3 - 4 \cdot x}$$

$$\frac{3}{4 \cdot x} - \frac{9}{8 \cdot (x - 2)} + \frac{11}{8 \cdot (x + 2)}$$

Exhibit 4-32

Laplace Transforms

1

$\dfrac{1}{s}$ So the Laplace Transform of 1 is $\dfrac{1}{s}$

t

$\dfrac{1}{s^2}$

$2 \cdot \sqrt{\dfrac{t}{\pi}}$

$\dfrac{1}{s^{\left(\frac{3}{2}\right)}}$

$\dfrac{\sin(a \cdot t)}{a}$

$\dfrac{1}{\left(s^2 + a^2\right)}$

Exhibit 4-33

Exhibit 4-33). For more information about using transforms in Mathcad, consult the Mathcad Users Guide and the Mathcad Treasury.

Evaluation style

Evaluation Style gives you control over how the results are displayed.

Show evaluation steps (radio buttons)

Vertically, inserting lines

Vertically, without inserting lines

Horizontally

If you use the right pointing arrow (the symbolic equal sign) to generate symbolic results, the answers will always be to the right of the arrow. However, the "Evaluation Style" choice on the Symbolics menu brings up a dialog box that allows you to control whether the answer is below the original expression (the default option), or to the right of the original expression. The option to insert lines between the steps of the derivation also exists.

Show comments (check box)

This check box allows you to turn on evaluation comments to describe the operations that were done on the original expression to get the given answer. Thus you can create a record of the steps in the evaluation process.

Evaluate in place (check box)

This overwrites the previous expressions when the symbolic evaluation takes place. This is useful when you want to use an expression in your work, but do not need to explain its origin to a reader. Derivation comments would normally not be appropriate here and are not an option with derive in place.

Other Symbolic Calculations

You can symbolically evaluate any of Mathcad's built-in functions. If Mathcad can compute an exact result, it will do so and return that result as the answer. Otherwise, Mathcad will return the original expression as the answer. In Exhibit 4-34 we see Mathcad's symbolic

---

$$\sum_{n=1}^{10} \frac{1}{n!}$$

$$\frac{6235301}{3628800}$$

$$\prod_{j=1}^{10} j^3$$

$$47784725839872000000$$

Exhibit 4-34

$$f(x) := x^2 \cdot \sin(x)$$

$$\frac{d}{dx} f(x) \rightarrow 2 \cdot x \cdot \sin(x) + x^2 \cdot \cos(x)$$

Now, if you edit the definition of the function, say to $x^5 \cdot \sin(x^2)$ Mathcad's live symbolics capability comes into play and automatically updates the result. In effect, the symbolic processor "knows" that the definition of f has been changed and responds to the change.

$$f(x) := x^5 \cdot \sin(x^2)$$

$$\frac{d}{dx} f(x) \rightarrow 5 \cdot x^4 \cdot \sin(x^2) + 2 \cdot x^6 \cdot \cos(x^2)$$

Exhibit 4-35

---

answers to a summation and an iterated product. These can be selected from the palettes or by keystroke combinations:

- Summation   [Ctrl][Shift]4
- Product   [Ctrl][Shift]3

Live Symbolics

Early in this book you saw that Mathcad's numerical calculations are live. Your results are automatically recalculated when you change a numerical value or edit a mathematical expression. Mathcad also has "live symbolics," which similarly revises your symbolic calculations. Suppose, for example, that you define a function and use it in a symbolic calculation. If you then edit the original problem, live symbolics will automatically update the answer. The symbolic processor "knows" the definition of the function and "understands" when you revise the problem that it should update the answer.

   In the example shown in Exhibit 4-35, we created a function and had Mathcad differentiate

---

Exhibit 4-36 Symbolic
Evaluation button.

Exhibit 4-37 Symbolic
Keyword Evaluation
button.

Exhibit 4-38 Floating
Point Evaluation
Keyword button.

Exhibit 4-39 Complex
Evaluation Keyword
button.

it symbolically using the right-pointing arrow operator to activate the symbolic process. We then edited the original function and Mathcad automatically calculated the symbolic derivative of the new function.

Now that you have seen how Mathcad does symbolic math, you are ready to look at a feature of Mathcad that is new in version 7: the "Symbolic Keyword" palette. You have already seen this palette in Chap. 1, when we looked at the main features of the Mathcad screen. Now we will look at the individual buttons on the Symbolic Keyword palette. When you first look at these buttons, you will find that most have names that will suggest their functions. You will probably be familiar with most of these key words from our perusal of the Symbolics menu and its submenus.

Exhibit 4-36 shows the Symbolic Evaluation button. This is the basic button, sometimes called the *symbolic equal sign,* that tells Mathcad to evaluate an expression symbolically. You can also invoke this command by typing a period while holding down the control key. Calculations made with the symbolic equal sign are live, just like Mathcad's numerical calculations. Calculations made with commands from the "Symbolics" menu are not live.

Exhibit 4-37 shows the "Symbolic Keyword Evaluation" button. This version of the symbolic equal sign allows you to specify additional information about your calculation, so it gives you two placeholders—one for the expression to be acted on, and one for the keyword that will tell Mathcad's symbolic processor how to customize the symbolic operation to be performed. If you want to use more than one keyword to customize the operation, you can invoke this button again and enter a new keyword.

Exhibit 4-38 shows the "Floating Point Evaluation Keyword" button. This produces a floating-point result with the number of decimal places you specify, by entering it in the placeholder.

The following buttons (see Exhibits 4-39 to 4-58), from the "Symbolic Keywords" palette, are used to initiate symbolic operations similar to ones you learned about earlier when we used the Symbolics menu. The names are very similar to those we saw before. You should have no difficulty using these buttons to perform the symbolic operations. An example is shown in Exhibit 4-23. Another example, to factor a number or expression:

- Type the number or expression.
- Press Ctrl+Shift+the period key, or use the second button on the Symbolic Keyword palette, the "Symbolic Keyword Evaluation" button (shown in Exhibit 4-37).
- Type "Factor" in the placeholder, or use the "Factor Expressions Keyword" button shown in Exhibit 4-49.

| expand | solve | simplify |

Exhibit 4-40 Expand Keyword button.

Exhibit 4-41 Solve for Variable Keyword button.

Exhibit 4-42 Simplify Expression Keyword button.

| substitute | collect | series |

Exhibit 4-43 Substitute Expression Keyword button.

Exhibit 4-44 Collect Terms Keyword button.

Exhibit 4-45 Expand in Series Keyword button.

| assume | parfrac | coeffs |

Exhibit 4-46 Assume Keyword button.

Exhibit 4-47 Convert to Partial Fraction Keyword button.

Exhibit 4-48 Polynomial Coefficients Keyword button.

The "Keyword Modifiers Palette" button (Exhibit 4-59) brings up a small palette that gives you additional control over the symbolic operations called forth by these symbolic keywords.

The "Assume Symbolic Modifier" button, in Exhibit 4-60, gives you the ability to introduce constraints to the calculation. To use a modifier in a symbolic simplification problem

- Enter the expression you want Mathcad to simplify.

- Press the "Ctrl+Shift+the period" key, or use the second button, the "Symbolic Keyword Evaluation" button (shown in Exhibit 4-37), on the "Symbolic Keyword" palette.

| factor | fourier | laplace |

Exhibit 4-49 Factor Expressions Keyword button.

Exhibit 4-50 Fourier Transform Keyword button.

Exhibit 4-51 Laplace Transform Keyword button.

| ztrans | invfourier | invlaplace |

Exhibit 4-52 Ztransform Keyword button.

Exhibit 4-53 Inverse Fourier Transform Keyword button.

Exhibit 4-54 Inverse Laplace Transform Keyword button.

invztrans

Exhibit 4-55 Inverse Ztransform Keyword button.

$M^T \rightarrow$

Exhibit 4-56 Symbolic Matrix Transpose Keyword button.

$M^{-1} \rightarrow$

Exhibit 4-57 Symbolic Matrix Inverse Keyword button.

$|M| \rightarrow$

Exhibit 4-58 Symbolic Matrix Determinant Keyword button.

Modifiers

Exhibit 4-59 Keyword Modifiers Palette button.

assume

Exhibit 4-60 Assume Symbolic Modifier button.

real

Exhibit 4-61 Real Symbolic Modifier button.

RealRange

Exhibit 4-62 RealRange Symbolic Modifier button.

trig

Exhibit 4-63 Trig Symbolic Modifier button.

Exhibit 4-61 shows the "Real Symbolic Modifier" button. This modifier tells Mathcad to assume that all indeterminate values are real as it simplifies the expression.

Exhibit 4-62 shows the "RealRange Symbolic Modifier" button. This modifier tells Mathcad to assume that all indeterminate values are real and between "a" and "b" as it simplifies the expression. Here, "a" and "b" are real numbers or infinity.

Exhibit 4-63 shows the "Trig Symbolic Modifier" button. This modifier uses the following two identities to simplify trigonometric expressions:

$$\sin(x)^2 + \cos(x)^2 = 1$$

$$\cosh(x)^2 - \sinh(x)^2 = 1$$

It does not simplify logs, powers, or radicals. To see a simple example of the use of "Trig," type in "sin(x)² + cos(x)²," then simplify, using the "Symbolic Keyword Evaluation" button shown in Exhibit 4-37 together with the "Trig" modifier. Mathcad will simplify the expression and return an answer of 1.

We have reached the end of this chapter on the use of the symbolic processor and derivatives, but it is important to understand that we have only scratched the surface of the ways in which Mathcad can help with symbolic mathematics. Mathcad "knows" the standard identities and mathematical relations and has a vast array of powerful techniques for solving equations and evaluating expressions. As you work with Mathcad, you will undoubtedly discover many additional ways to use these capabilities in your own work.

Trigonometry, Hyperbolic Functions, and Some Applications to Astronomy

Introduction

In this chapter we see how Mathcad makes working with trigonometric and hyperbolic functions, and their inverses, easy. Mathcad makes these functions instantly available when you type in the function name—or select them from the "Function" choice on the "Insert" menu. This will be described in greater detail in Chap. 10. There is no need to use tables or even scientific calculators. Mathcad makes all the data you need for such calculations readily available and easy for you to use. Later in this chapter we look at some examples of astronomy problems, many of which involve the use of trigonometric functions.

Trigonometric Functions

Mathcad puts the full complement of trigonometric functions at your fingertips. When given an angle in radians as an argument, it presents the value of the function instantly. The following table summarizes the available functions.

In Exhibit 5-1 we see examples of these trigonometric functions.

Mathcad's trigonometric functions take arguments that are in radians. If you are working with degrees, you will need to convert to radians before passing the angle to Mathcad to compute the trigonometric function. Fortunately, Mathcad provides an easy way to make this conversion. As you type the argument for a trigonometric function in degrees, just multiply it by "deg" (i.e., type "*deg"). For example, if you want Mathcad to compute the sine of 37°, you would type "sin(37*deg)." See some examples in Exhibit 5-2. Also, although it is not the recommended method, if you prefer to work from what might be called "first principles" instead of letting Mathcad convert the angle for you, you can multiply the angle in degrees by "2*pi/ 360" or what amounts to the same thing "π/180" before passing it to a Mathcad trigonometry function (see Exhibit 5-3). The value of the trigonometric function is returned by Mathcad.

Type	Returns	Comment
sin(x)	Sine	In a right triangle, the sine is the ratio of the length of the side opposite the angle x divided by the length of the hypotenuse.
cos(x)	Cosine	In a right triangle, the cosine is the ratio of the length of the side adjacent to angle x divided by the length of the hypotenuse.
tan(x)	Tangent	In a right triangle, the tangent is the ratio of the length of the side opposite angle x divided by the length of the adjacent side [or, equivalently, it equals sin(x)/cos(x)]. The angle x should not be an odd multiple of π/2, because the tangent becomes infinite for such values.
cot(x)	Cotangent	In a right triangle, the cotangent is the ratio of the length of the side adjacent angle x divided by the length of the opposite side [or, equivalently, it equals cos(x)/sin(x) or 1/tan(x)]. The angle x should not be an integer multiple of π.
sec(x)	Secant	In a right triangle, the secant is the ratio of the length of the hypotenuse divided by the length of the adjacent side to angle x [or, equivalently, it equals 1/cos(x)]. The angle x should not be an odd multiple of π/2.
csc(x)	Cosecant	In a right triangle, the cosecant is the ratio of the length of the hypotenuse divided by the length of the opposite side to angle x [or, equivalently, it equals 1/sin(x)]. The angle x should not be an integer multiple of π.

Mathcad's trigonometric functions also take complex numbers as arguments (see Exhibit 5-4) and can be used with the symbolic processor (see Exhibit 5-5).

Inverse trigonometric functions

Mathcad provides three inverse trigonometric functions. These take as an argument a number that represents the value of one of the three functions (sine, cosine, or tangent) of an angle. These functions return the angle which has the given value of its trigonometric function. The angle returned by Mathcad will be in radians and will be

- Between $-\pi/2$ and $\pi/2$ for arcsine and arctangent
- Between 0 and π for arccosine

If you want the angle to be in degrees, divide by "deg" or type "deg" in the placeholder when Mathcad returns the value of the angle (in radians). You can also divide by $\pi/180$ (or multiply by $180/\pi$). Some examples are shown in Exhibit 5-6. Use the following commands to compute inverse trigonometric functions.

Type	Returns	Comment
asin(x)	Inverse sine or arcsine	Returns the angle that has its sine equal to x
acos(x)	Inverse cosine or arc cosine	Returns the angle that has its cosine equal to x
atan(x)	Inverse tangent or arc tangent	Returns the angle that has its tangent equal to x

$\sin(.55) = 0.52269$

$\cos(.55) = 0.853$

$\tan(.55) = 0.613$

$\cot(.55) = 1.631$

$\sec(.55) = 1.173$

$\csc(.55) = 1.913$

$i := 1 .. 8$

$a_i := \sin\left(\dfrac{i}{10}\right)$

$$a = \begin{bmatrix} 0 \\ 0.1 \\ 0.199 \\ 0.296 \\ 0.389 \\ 0.479 \\ 0.565 \\ 0.644 \\ 0.717 \end{bmatrix}$$

$b_i := \cos\left(\dfrac{i}{10}\right)$

$$b = \begin{bmatrix} 0 \\ 0.995 \\ 0.98 \\ 0.955 \\ 0.921 \\ 0.878 \\ 0.825 \\ 0.765 \\ 0.697 \end{bmatrix}$$

$c_i := \tan\left(\dfrac{i}{10}\right)$

$$c = \begin{bmatrix} 0 \\ 0.1 \\ 0.203 \\ 0.309 \\ 0.423 \\ 0.546 \\ 0.684 \\ 0.842 \\ 1.03 \end{bmatrix}$$

$d_i := \cot\left(\dfrac{i}{10}\right)$

$$d = \begin{bmatrix} 0 \\ 9.967 \\ 4.933 \\ 3.233 \\ 2.365 \\ 1.83 \\ 1.462 \\ 1.187 \\ 0.971 \end{bmatrix}$$

$e_i := \sec\left(\dfrac{i}{10}\right)$

$$e = \begin{bmatrix} 0 \\ 1.005 \\ 1.02 \\ 1.047 \\ 1.086 \\ 1.139 \\ 1.212 \\ 1.307 \\ 1.435 \end{bmatrix}$$

$f_i := \csc\left(\dfrac{i}{10}\right)$

$$f = \begin{bmatrix} 0 \\ 10.017 \\ 5.033 \\ 3.384 \\ 2.568 \\ 2.086 \\ 1.771 \\ 1.552 \\ 1.394 \end{bmatrix}$$

Exhibit 5-1

45 degrees equals 45*2*pi/360 radians

$$45 \cdot 2 \cdot \frac{\pi}{360} = 0.785$$

$\sin(0.785) = 0.707$

$\sin(45 \cdot \deg) = 0.707$

$\sin(45.785 \cdot \deg) = 0.717$

$$\sin\left(45 \cdot 2 \cdot \frac{\pi}{360}\right) = 0.707$$

$\sin(2 \cdot i \ ) = 3.627i$

Exhibit 5-2   **Exhibit 5-3**   **Exhibit 5-4**

If your work requires you to have angles expressed in degrees, minutes, and seconds, you will find a description in Chap. 6 of how to convert angles from degrees and decimal degrees into degrees, minutes, and seconds.

Angles subtended by sun and moon

The distance between the moon and the earth varies over a range of about 11 percent, but averages about 384,000 km. The radius of the moon is about 1738 km (so the diameter =

$$\operatorname{asin}\left(\frac{\sqrt{2}}{2}\right) = 0.785$$

To get result in degrees, divide by "deg"

$$\frac{\operatorname{asin}\left(\frac{\sqrt{2}}{2}\right)}{\deg} = 45$$

Or type "deg" in placeholder

$\sin\left(\frac{\pi}{4}\right)$

$$\operatorname{asin}\left(\frac{\sqrt{2}}{2}\right) = 0.785$$

$\frac{1}{2} \cdot \sqrt{2}$

$$\operatorname{asin}\left(\frac{\sqrt{2}}{2}\right) = 45 \cdot \deg$$

Moon

$$\operatorname{atan}\left(\frac{1738 \cdot 2}{384400}\right) = 9.042417 \cdot 10^{-3}$$

$$\frac{\operatorname{atan}\left(\frac{1738 \cdot 2}{384400}\right)}{\deg} = 0.518092$$

Sun

$$\operatorname{atan}\left(\frac{1392000}{149600000}\right) = 9.304544 \cdot 10^{-3}$$

$$\frac{\operatorname{atan}\left(\frac{1392000}{149600000}\right)}{\deg} = 0.533111$$

Exhibit 5-5   **Exhibit 5-6**   **Exhibit 5-7**

2*1738). If you imagine that the diameter of the moon is the opposite side of a right triangle and the average distance between the earth and the moon is the adjacent side of the triangle, the tangent of the angle would be 2*1738/384,000 and the angle would be found by computing the arctangent. You will notice that we are ignoring the fact than an observer would be on the surface of the earth, not at the center, and so a slightly different formula would be required.

Similarly for the sun, the diameter is 1,392,000 km and the average distance from the earth is 149,600,000 km. By a similar process, we compute the angular size of the sun in the sky (see Exhibit 5-7). It is interesting to note that the average angular sizes of the sun and moon are close to being equal. This rather remarkable coincidence is what accounts for the existence of solar eclipses, when the moon blocks the sun's light for a time. The distances of both the sun and moon from the earth are variable over a few percent. As a result, the amount of the sun that is blocked by the moon during an eclipse varies slightly from eclipse to eclipse.

Another commonly observed phenomenon related to the angular size of heavenly bodies is the distortion of the apparent size of an object, such as the sun, resulting from the diffraction of light from its surface as it passes through the earth's atmosphere. It is more apparent when the object appears to be near the horizon than when it is high overhead, near the zenith.

Trigonometric identities

There are many trigonometric identities (equations that involve trigonometric functions that are valid for all values of the functions). Working with these is often required in many fields such as mathematics, engineering, and surveying. Exhibit 5-8 gives some examples of how Mathcad can help you work with these identities.

In Exhibit 5-9, we also see an example of a helpful feature of Mathcad's Professional version, called a *QuickSheet,* which is an on-line template for work that has proved to be frequently needed by power users of Mathcad. As a result, Mathcad has added QuickSheets in a manner that enables users of the Professional version to access, use and incorporate them into their Mathcad documents. The QuickSheet in this exhibit computes all the trigonometric functions with a given argument and then provides a plot of the six trigonometric functions.

Hyperbolic Functions

Mathcad gives you the full range of hyperbolic functions, summarized in the following table. Some examples of the use of these functions are seen in Exhibit 5-10.

Type	Returns	Comment
$\sinh(x)$	Hyperbolic sine	Returns the hyperbolic sine of x, which is equal to $(e^x - e^{-x})/2$
$\cosh(x)$	Hyperbolic cosine	Returns the hyperbolic cosine of x, which is equal to $(e^x + e^{-x})/2$
$\tanh(x)$	Hyperbolic tangent	Returns the hyperbolic cotangent of x, which is equal to $\sinh(x)/\cosh(x)$
$\coth(x)$	Hyperbolic cotangent	Returns the hyperbolic tangent of x, which is equal to $1/\tanh(x)$
$\text{sech}(x)$	Hyperbolic secant	Returns the hyperbolic secant of x, which is equal to $1/\cosh(x)$
$\text{csch}(x)$	Hyperbolic cosecant	Returns the hyperbolic cosecant of x, which is equal to $1/\sinh(x)$

$x := .2$

$y := 1.2$

Trigonometric Identity	Verification

$\sin(x)^2 + \cos(x)^2 = 1$

$1 + \tan(x)^2 = \sec(x)^2$

$\sin(2 \cdot x) = 2 \cdot \sin(x) \cdot \cos(x)$

$\cos(2 \cdot x) = \cos(x)^2 - \sin(x)^2$

Or

$\cos(2 \cdot x) = 2 \cdot \cos(x)^2 - 1$

Or

$\cos(2 \cdot x) = 1 - 2 \cdot \sin(x)^2$

$\sin(x + y) = \sin(x) \cdot \cos(y) + \cos(x) \cdot \sin(y)$

Or

$\sin(x - y) =$

Verification column:

$\sin(x)^2 + \cos(x)^2 = 1$

$1 + \tan(x)^2 - \sec(x)^2 = 0$

$\sin(2 \cdot x) - 2 \cdot \sin(x) \cdot \cos(x) = 0$

$\cos(2 \cdot x) - \cos(x)^2 + \sin(x)^2 = 0$

$\cos(2 \cdot x) - 2 \cdot \cos(x)^2 + 1 = 0$

$\cos(2 \cdot x) - 1 + 2 \cdot \sin(x)^2 = 0$

$\sin(x + y) - \sin(x) \cdot \cos(y) - \cos(x) \cdot \sin(y) = 0$

Or, choosing the minus sign

$\sin(x - y) - \sin(x) \cdot \cos(y) + \cos(x) \cdot \sin(y) = 0$

Mathcad can do more than just verify trigonometric identities that you enter. The symbolic processor "knows" many identities and will provide them if you enter the left hand side of the equation and then use the symbolic menu to simplify or expand. For example

$\sin(x)^2 + \cos(x)^2$

1

Or, second example

$1 + \tan(x)^2$

$\dfrac{1}{\cos(x)^2}$ Which is equivalent to $\sec(x)^2$

Exhibit 5-8

Third example

$\sin(2 \cdot x)$

$2 \cdot \sin(x) \cdot \cos(x)$

Fourth example

$\cos(2 \cdot x)$

$2 \cdot \cos(x)^2 - 1$

Fifth example

$\sin(x + y)$

$\sin(x) \cdot \cos(y) + \cos(x) \cdot \sin(y)$

Sixth example

$\sin(x - y)$

$\sin(x) \cdot \cos(y) - \cos(x) \cdot \sin(y)$

Exhibit 5-8 (*Continued*)

As we saw with trigonometric functions, hyperbolic functions can take complex or imaginary arguments and can be used with the symbolic processor. See Exhibit 5-11 for examples.

Relationship between trigonometric and hyperbolic functions

There are many analogies between trigonometric and hyperbolic functions. Some examples are demonstrated in Exhibit 5-12.

Inverse hyperbolic functions

Type	Returns	Comment
asinh(x)	Inverse hyperbolic sine	Returns the number that has its hyperbolic sine equal to x
acosh(x)	Inverse hyperbolic cosine	Returns the number that has its hyperbolic cosine equal to x
atanh(x)	Inverse hyperbolic tangent	Returns the number that has its hyperbolic tangent equal to x

See Exhibit 5-13.

Trigonometric Functions (Arguments in Radians)

(From Mathcad Quick Sheets)

Calculates values for the basic trigonometric functions for an argument given in radians.

Enter a number: $x := 3.34$

Values of trig functions at x:

$\sin(x) = -0.197$ $\csc(x) = -5.073$

$\cos(x) = -0.98$ $\sec(x) = -1.02$

$\tan(x) = 0.201$ $\cot(x) = 4.974$

$r := -2 \cdot \pi, -2 \cdot \pi + 0.025 .. 2 \cdot \pi$

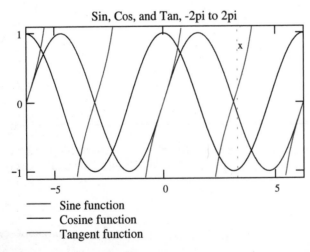

Exhibit 5-9 Example of a QuickSheet.

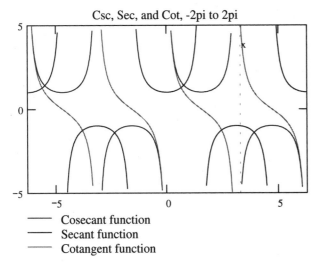

Csc, Sec, and Cot, -2pi to 2pi

—— Cosecant function
—— Secant function
—— Cotangent function

Exhibit 5-9 *(Continued)*

Relationship between trigonometric identities and identities involving hyperbolic functions

Many trigonometric identities have analogies in equations that involve hyperbolic functions, as you can see in Exhibit 5-14.

For derivatives of trigonometric and hyperbolic functions, see Exhibit 5-15; for series representation of hyperbolic functions, see Exhibit 5-16.

Parachute problem

Hyperbolic functions frequently arise in the solutions of differential equations and often are used in the mathematical representation of hanging cables. This problem involves a falling parachute holding a mass "M" and with air resistance proportional to the square of the velocity (see Exhibit 5-17). The coefficient "k" is related to the shape of the parachute. Sokolnikoff and Sokolnikoff work out an analytical solution to this example in their book *Higher Math for Engineers and Physicists* (McGraw-Hill, 1941). Here, we show how to use Mathcad to compute numerical results.

Converting standard dates to Julian days

For many types of work it is necessary to deal with dates and time intervals. For interest rate and other date-sensitive calculations, spreadsheets often use a system of counting days (commonly starting around the beginning of 1900). They assign numbers to days, and fractions of days, and then compute intervals of time by subtracting. Many astronomical problems involve observations of celestial events that take place over a long period of time and require an accurate calculation of the time interval. To facilitate such calculations, astronomers use a system of counting days called the *Julian dates.* The system was proposed in 1583 by the French classical scholar Joseph Scaliger and named in honor of his father, Julius Caesar Scaliger. In the Julian day system (or Julian dates), days are numbered consecutively, starting from time 0.0 at Greenwich mean noon of January 1, 4713 B.C. The starting time of noon GMT [Greenwich mean time; also called *Universal Time* (UT)] is convenient for many optical astronomers, but for many nonastronomical purposes a modified version of the Julian date is also used. The modified Julian date is defined as equal to the Julian date minus 2,400,000.5. This modified Julian date begins at midnight (instead of noon) and for dates in the twentieth

$\sinh(.55) = 0.57815$

$\cosh(.55) = 1.155$

$\tanh(.55) = 0.501$

$\coth(.55) = 1.998$

$\text{sech}(.55) = 0.866$

$\text{csch}(.55) = 1.73$

$i := 1 .. 8$

$$a_i := \sinh\left(\frac{i}{10}\right) \qquad b_i := \cosh\left(\frac{i}{10}\right) \qquad c_i := \tanh\left(\frac{i}{10}\right)$$

$$a = \begin{bmatrix} 0 \\ 0.1 \\ 0.201 \\ 0.305 \\ 0.411 \\ 0.521 \\ 0.637 \\ 0.759 \\ 0.888 \end{bmatrix} \qquad b = \begin{bmatrix} 0 \\ 1.005 \\ 1.02 \\ 1.045 \\ 1.081 \\ 1.128 \\ 1.185 \\ 1.255 \\ 1.337 \end{bmatrix} \qquad c = \begin{bmatrix} 0 \\ 0.1 \\ 0.197 \\ 0.291 \\ 0.38 \\ 0.462 \\ 0.537 \\ 0.604 \\ 0.664 \end{bmatrix}$$

Exhibit 5-10

$$d_i := \coth\left(\frac{i}{10}\right) \qquad d = \begin{bmatrix} 0 \\ 10.033 \\ 5.066 \\ 3.433 \\ 2.632 \\ 2.164 \\ 1.862 \\ 1.655 \\ 1.506 \end{bmatrix}$$

$$e_i := \operatorname{sech}\left(\frac{i}{10}\right) \qquad e = \begin{bmatrix} 0 \\ 0.995 \\ 0.98 \\ 0.957 \\ 0.925 \\ 0.887 \\ 0.844 \\ 0.797 \\ 0.748 \end{bmatrix}$$

$$f_i := \operatorname{csch}\left(\frac{i}{10}\right) \qquad f = \begin{bmatrix} 0 \\ 9.983 \\ 4.967 \\ 3.284 \\ 2.435 \\ 1.919 \\ 1.571 \\ 1.318 \\ 1.126 \end{bmatrix}$$

Exhibit 5-10 (*Continued*)

$x := .5$

$\sin(x)$

$S := \sin(x)$

$S = 0.479$

$\sinh(x)$

$SH := \dfrac{\sinh(i \cdot x)}{i}$

$SH = 0.479$

$\cos(x)$

Complex and Imaginary

$C := \cos(x)$

$C = 0.878$

$\sinh(1 + .25 \cdot i) = 1.138667 + 0.381764i$

$\cosh(x)$

$\cosh(1 + .25 \cdot i) = 1.495 + 0.291i$

$CH := \cosh(i \cdot x)$

$\tanh(1 + .25 \cdot i) = 0.782 + 0.103i$

$CH = 0.878$

$\dfrac{1.138667 + 0.381764i}{1.49511 + 0.290749i} = 0.782 + 0.103i$

$\tan(x)$

$T := \tan(x)$

$T = 0.546$

Symbolic

$\tanh(x)$

$\dfrac{\sinh(x)}{\cosh(x)} - \tanh(x)$

$TH := \dfrac{\sinh(i \cdot x)}{i \cdot \cosh(i \cdot x)}$

0

$TH = 0.546$

Exhibit 5-11

Exhibit 5-12 Relationships
between trigonometric
and hyperbolic
functions.

$asinh(.35) = 0.343222$

$acosh(.35) = 1.213i$

$atanh(.35) = 0.365$

Check

$\sinh(0.343222) = 0.35$

$\cosh(1.213225i) = 0.35$

$\tanh(0.365444) = 0.35$

Exhibit 5-13 Inverse
hyperbolic functions.

$x := .2$ $y := .1$

Hyperbolic Identity	Verification

$\cosh(x)^2 - \sinh(x)^2 = 1$

$\cosh(x)^2 - \sinh(x)^2 = 1$

$1 - \tanh(x)^2 = \mathrm{sech}(x)^2$

$1 - \tanh(x)^2 - \mathrm{sech}(x)^2 = 0$

$\sinh(2 \cdot x) = 2 \cdot \sinh(x) \cdot \cosh(x)$

$\sinh(2 \cdot x) - 2 \cdot \sinh(x) \cdot \cosh(x) = 0$

$\cosh(2 \cdot x) = \cosh(x)^2 - \sinh(x)^2$

$\cosh(2 \cdot x) - \cosh(x)^2 - \sinh(x)^2 = 0$

Or

$\cosh(2 \cdot x) = 2 \cdot \cosh(x)^2 - 1$

$\cosh(2 \cdot x) - 2 \cdot \cosh(x)^2 + 1 = 0$

$\sinh(x + y) = \sinh(x) \cdot \cosh(y) + \cosh(x) \cdot \sinh(y)$

$\sinh(x + y) - \sinh(x) \cdot \cosh(y) - \cosh(x) \cdot \sinh(y) = 0$

Or   Or, choosing the minus sign

$\sinh(x - y) = \sinh(x) \cdot \cosh(y) - \cosh(x) \cdot \sinh(y)$

$\sinh(x - y) - \sinh(x) \cdot \cosh(y) + \cosh(x) \cdot \sinh(y) = 0$

Mathcad can do more than just verify hyperbolic identities that you enter. The symbolic processor "knows" many identities and will provide them if you enter the left hand side of the equation and then use the symbolic menu to simplify or expand. For example

$\cosh(x)^2 - \sinh(x)^2$

1

Or, second example

$1 - \tanh(x)^2$

$\dfrac{1}{\cosh(x)^2}$ Which is equivalent to $\mathrm{sech}(x)^2$.

Exhibit 5-14 Relationship between trigonometric identities and identities involving hyperbolic functions.

Third example

$\sinh(2\cdot x)$

$2\cdot\sinh(x)\cdot\cosh(x)$

Fourth example

$\cosh(2\cdot x)$

$2\cdot\cosh(x)^2 - 1$

Fifth example

$\sinh(x + y)$

$\sinh(x)\cdot\cosh(y) + \cosh(x)\cdot\sinh(y)$

Sixth example

$\sinh(x - y)$

$\sinh(x)\cdot\cosh(y) - \cosh(x)\cdot\sinh(y)$

Exhibit 5-14 (*Continued*)

---

or twenty-first century, has much smaller, more convenient, numbers. The starting point for modified Julian dates is November 17, 1858. For example

- The Greenwich Mean Time (GMT) noon of November 14, 1981 was Julian date (JD) 2,444,923.0, and the preceding midnight was at JD 2,444,922.5
- The modified Julian date (MJD) for November 14, 1981 began at midnight instead of noon and so GMT noon of 11-14-81 was 2,444,923.0 minus 2,400,000.5 or 44,922.5
- The preceding midnight was at JD 2,444,922.5. The corresponding MJD was 2,444,922.5 minus 2,400,000.5 or 44,922.0

As you can see, the fractional part of a day is represented by a decimal fraction, so any point in time can be represented by a Julian day, including the decimal part of the day, and the

$x := .5$

$\sinh(x)$

$$S := x + \sum_{n=2}^{80} \frac{x^{2 \cdot n - 1}}{2 \cdot n - 1\,!}$$

$S = 0.521$

As compared to

$\sinh(.5) = 0.521$

$\dfrac{d}{dx}\sin(x)$

$\cos(x)$

$\dfrac{d}{dx}\cos(x)$

$-\sin(x)$

$\dfrac{d}{dx}\tan(x)$

$1 + \tan(x)^2$

Using the identity $1 + \tan(x)^2 = \sec(x)^2$ this equals $\sec(x)^2$

$\dfrac{d}{dx}\sinh(x)$

$\cosh(x)$

$\dfrac{d}{dx}\cosh(x)$

$\sinh(x)$

$\dfrac{d}{dx}\tanh(x)$

$1 - \tanh(x)^2$

$1 - \tanh(x)^2 = \operatorname{sech}(x)^2$

So this equals $\operatorname{sech}(x)^2$

$\cosh(x)$

$$C := 1 + \sum_{n=1}^{80} \frac{x^{2 \cdot n}}{2 \cdot n\,!}$$

$C = 1.128$

As compared to

$\cosh(.5) = 1.128$

Exhibit 5-15 Derivatives of trigonometric and hyperbolic functions.

Exhibit 5-16 Series representation of hyperbolic functions.

interval between two events can then be precisely determined by a subtraction. For example, in Central Standard Time (CST), a new Julian day begins at 6:00 A.M. CST (because it is 12:00 noon GMT or UT). So for 11:03 P.M. CST on January 7, 1996, the Julian date is 2,450,090.710.

In Exhibit 5-18 we see how dates in a standard calendar can be converted to Julian dates. This and the two subsequent exhibits deal with the use of Julian dates. These exhibits are based on algorithms from *Practical Astronomy with Your Calculator* by Peter Duffett-Smith (Cambridge University Press, 1981). Similar algorithms are available in other books of computational astronomy.

The calculations involving Julian dates require the use of some elementary Mathcad programming. Programming is discussed more fully in Chap. 10. For these exhibits, it is sufficient

According to Sokolnikoff and Sokolnikoff, if a parachute is supporting a mass "M"
and falling toward the ground, and if air resistance is proportional
to the square of the velocity, then

the velocity "v" is given by the following formula.

"k" is a constant based on the design of the parachute
"t" is the time
"g" is the acceleration of gravity

$$k := .5 \cdot \frac{kg}{m}$$

$$t := 8 \cdot sec$$

$$g := 9.8 \cdot \frac{m}{sec^2}$$

$$M := 1 \cdot kg$$

$$v := \sqrt{g \cdot \frac{M}{k} \left[\tanh \left[\left(\sqrt{k \cdot \frac{g}{M}} \right) \cdot t \right] \right]}$$

$$v = 4.427 \bullet m \bullet s^{-1}$$

Also, the distance fallen is

$$s := \frac{M}{k} \cdot \ln \left(\cosh \left(\sqrt{\frac{k \cdot g}{M}} \cdot t \right) \right)$$

$$s = 34.03 \bullet m$$

Exhibit 5-17 Falling-parachute problem illustrating hyperbolic function.

To convert a date into a Julian Day number

Example: 6 A.M. on February 17, 1985

$y := 1985$

$m := 2$

$d := 17.25$

Step 1

If m=1 or m=2, subtract 1 from y and add 12 to m
Otherwise, continue

In our example, y becomes 1984, and m becomes 2+12=14

$y := if((m \equiv 1) + (m \equiv 2), y - 1, y)$

$m := if((m \equiv 1) + (m \equiv 2), m + 12, m)$

Step 2

If the date is later than October 15, 1582, calculate

$$A = \text{integer part of } \frac{y}{100}$$

$$A := floor\left(\frac{y}{100}\right)$$

$$B = 2 - A + \text{integer part of } \frac{A}{4}$$

Otherwise, B=0

$$A := if\left[(y > 1582) + ((y \equiv 1582) \cdot (m > 10)) + ((y \equiv 1582) \cdot (m \equiv 10) \cdot (d > 15)), floor\left(\frac{y}{100}\right), floor\left(\frac{y}{100}\right)\right]$$

$A = 19$

$$B := if\left[(y > 1582) + ((y \equiv 1582) \cdot (m > 10)) + ((y \equiv 1582) \cdot (m \equiv 10) \cdot (d > 15)), 2 - A + floor\left(\frac{A}{4}\right), 0\right]$$

$B = -13$

Step 3

Calculate C = the integer part of $365.25 \cdot y$

$C := floor(365.25 \cdot y)$

$C = 724656$

Exhibit 5-18 Converting dates in a standard calendar to Julian dates.

Step 4

Calculate D = the integer part of $30.6001 \cdot (m + 1)$

$D := \text{floor}(30.6001 \cdot (m + 1))$

$D = 459$

Then the Julian Day is $B + C + D + d + 1720994.5$

$JD := B + C + D + d + 1720994.5$

$JD = 2446113.75$

Modified Julian date is $JD - 2400000.5$

$MJD := JD - 2400000.5$

$MJD = 46113.3$

Exhibit 5-18 (*Continued*)

---

to be aware that Mathcad allows you to create complicated conditional statements using boolean operators. For example, "and" gates are created using multiplication and "or" gates using addition.

Converting Julian days to dates

In Exhibit 5-19 we see how to convert Julian dates to standard calendar dates.

Converting Julian days to days of the week

In Exhibit 5-20 we see how to determine the day of the week if we have a Julian day.

Sidereal time

Sidereal time is of crucial importance to astronomers in determining the positions of astronomical objects in the sky. In Exhibit 5-21, we see an example of how to compute the sidereal time, given the Greenwich Mean Time (or Universal Time). This example is based on an algorithm from Mathcad's electronic book, *Astronomical Formulas,* by Martin Zombeck (Math Soft Inc., 1993). A complete description of how Mathcad can be used for problems involving astronomical calculations is beyond the scope of this book, but this example should help you see another area where Mathcad can be of great assistance to you in your work. The electronic book on astronomy is full of helpful methods that will put Mathcad to work for you on astronomical calculations.

To convert a Julian Day number to a date

Example: 2446114.25

$$JD := 2446114.25$$

$$JD = 2446114.25$$

Set the integral part = I and the fractional part = F

$$I := floor(JD) \qquad I = 2446114$$

$$F := JD - floor(JD) \qquad F = 0.25$$

Calculate A,

If I is greater than 2299160 A= integer part of $\dfrac{I - 1867216.25}{36524.25}$

Otherwise, set A = I

$$A := if\left(I > 2299160, floor\left(\frac{I - 1867216.25}{36524.25}\right), I\right)$$

$$A = 15$$

$$B := I + 1 + A - floor\left(\frac{A}{4}\right)$$

$$B = 2446127.0$$

Next, calculate C = B+ 1524 and D = integer part of $\dfrac{C - 122.1}{365.25}$

$$C := B + 1524$$

$$C = 2447651.0$$

$$D := floor\left(\frac{C - 122.1}{365.25}\right)$$

$$D = 6700.0$$

Calculate E = integer part of $365.25 \cdot D$ and G = integer part of $\dfrac{C - E}{30.6001}$

$$E := floor(365.25 \cdot D)$$

$$E = 2447175.0$$

$$G := floor\left(\frac{C - E}{30.6001}\right)$$

$$G = 15.0$$

Exhibit 5-19 Converting days in a standard calendar to Julian days.

Calculate the day of the month, d from C-E+F-integer part of $30.6001 \cdot G$

This is the day of the month plus the decimal fraction of a day

$$d := C - E + F - floor(30.6001 \cdot G)$$

Calculate the month,
If the integer G is less than 13.5, set m = G -1

Otherwise, set m = G-13

$$m := if(G < 13.5, G - 1, G - 13)$$

Calculate the year,
If m is more than 2.5, y = D - 4716

Otherwise, use y = D - 4715

$$y := if(m > 2.5, D - 4716, D - 4715)$$

So the date is

$m = 2$ February

$d = 17.25$ 6 p.m. on the 17th at GMT

$y = 1985$

Exhibit 5-19 (*Continued*)

Calculate A = (JD + 1.5) / 7

Take the fractional part of A
multiply by 7
and round to the nearest integer

if this number is	0, the day is	Sunday
	1	Monday
	2	Tuesday
	3	Wednesday
	4	Thursday
	5	Friday
	6	Saturday

Example

JD = 2446113.5

JD := 2446113.5

$$A := \frac{JD + 1.5}{7} \qquad A = 349445.0000$$

DD := A − floor(A)

DD·7 = 0   So this day is a Sunday

Exhibit 5-20 Converting Julian days to days of the week.

If you work with astronomy, you might also want to direct your attention to the World Wide Web site of the Directorate of Time at the U.S. Naval Observatory:

http://tycho.usno.navy.mil/estclock.html

This site provides a wealth of useful information and current quotes of time, including your local standard time and sidereal time. It also has a link to the Web site of the Greenwich Observatory (in Greenwich, England).

Coordinate Conversions

A common and very practical use of Mathcad is for conversions of coordinates. This is one of those often needed and straightforward, but tedious, chores, which people are bad at and

Universal to Sidereal Time Conversion

This document is used to convert any instant of **Universal Time** into **sidereal time.**

Sidereal time is based upon the Earth's rotation with respect to the fixed stars. **Local Sidereal Time** (LST) is the hour angle of the **vernal equinox**. Greenwich Sidereal Time (GST) is the local sidereal time at the Greenwich meridian. Universal time (UT) is the basis for civil timekeeping. It is closely related to the mean diurnal motion of the Sun and is directly related to sidereal time.

Local Apparent Sidereal Time (LAST) is the hour angle of the **true equinox** - the intersection of the true equator and ecliptic of the date - which is affected by the **nutation** of the Earth's axis. Local Mean Sidereal Time (LMST) is the hour angle of the mean **vernal equinox** - the intersection of the mean equator with the ecliptic of the date - which is affected by the **precession** of the Earth's axis. The relationship between the Local and Greenwich sidereal times are

LAST = GAST - L

LMST = GMST - L

where L is the observer's geographical **longitude.**

Relationship Between Various Sidereal Times

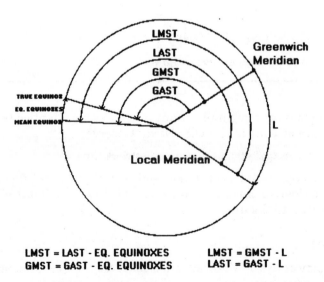

LMST = LAST - EQ. EQUINOXES   LMST = GMST - L
GMST = GAST - EQ. EQUINOXES   LAST = GAST - L

Exhibit 5-21 Converting UT to sidereal time.

Input Data:

Year: $Y := 1987$

Month: $M := 4$

Day of Month: $D := 10$

Time of Day (UT):

$$\begin{bmatrix} d_{hr} & d_{min} & d_{sec} \end{bmatrix} := (19 \quad 21 \quad 0.0)$$

Observer's longitude (positive, if west of Greenwich:

$$\begin{bmatrix} Lg_{deg} & Lg_{min} & Lg_{sec} \end{bmatrix} := (77 \quad 03 \quad 56)$$

Observer's location: U.S. Naval Obs., Wash., D.C.

Results:

Convert to the Sexagesimal System

Greenwich Mean Sidereal Time for the instant:

$GMST := dec2sexa(GMST)$

$GMST := (8 \quad 34 \quad 57.09)$

Local Mean Sidereal Time for the instant:

$LMST := dec2sexa(LMST)$

$LMST := (3 \quad 26 \quad 41.36)$

Greenwich Apparent Sidereal Time for the instant:

$GAST := dec2sexa(GAST)$

$GAST := (8 \quad 34 \quad 56.85)$

Local Apparent Sidereal Time for the instant:

$LAST := dec2sexa(LAST)$

$LAST := (3 \quad 26 \quad 41.11)$

Exhibit 5-21 (Continued)

computers are well suited for. You could get a handbook of mathematics and use the formulas for coordinate transformation to set up Mathcad procedures for such conversion, but Mathcad has already created excellent ready-to-use routines for these problems. They are available from MathSoft Inc., Mathcad's developer, in an electronic book called *Advanced Math* (1993). In addition to many other topics in advanced mathematics, the electronic book includes conversions for

- Rectangular to cylindrical
- Cylindrical to rectangular
- Rectangular to spherical
- Spherical to rectangular
- Cylindrical to spherical
- Spherical to cylindrical

In Exhibits 5-22 to 5-26 we illustrate the use of the 3D rectangular-to-spherical and spherical-to-rectangular conversion routines. You simply input the three rectangular coordinates or the three spherical coordinates, and Mathcad converts them for you. We also show examples of 2D rectangular-to-polar and polar-to-rectangular conversions in Exhibits 5-25 and 5-26.

The three-dimensional conversion functions use the following coordinate definitions to define a point X:

Rectangular: **Cylindrical:** **Spherical:**

(x, y, z)   (r, ϕ, z)   (r, ϕ, θ)

- ϕ is an angle with the positive x-axis, between 0 and 2π

- θ is an angle with the positive z-axis, between 0 and π

Exhibit 5-22

Coordinate Transformations

From the electronic book on Advanced Mathematics

Rectangular to Spherical: **(x, y, z)** $\longrightarrow$ **(r, ϕ, θ)**

$$m(X) := X \cdot \overrightarrow{\begin{bmatrix} 1 \\ 1 \\ 0 \end{bmatrix}}$$

$$RtoS(X, measure) := \begin{bmatrix} |X| \\ if\left(|m(X)|, \dfrac{angle(X_0, X_1)}{measure}, 0\right) \\ if\left(|X|, \dfrac{acos\left(\dfrac{X_2}{|X|}\right)}{measure}, 0\right) \end{bmatrix}$$

Example:

$$A := \begin{bmatrix} 1 \\ -1 \\ -10 \end{bmatrix} \qquad RtoS(A, deg) = \begin{bmatrix} 10.1 \\ 315 \\ 171.951 \end{bmatrix}$$

Or

$$B := \begin{bmatrix} 13 \\ -1 \\ -9 \end{bmatrix} \qquad RtoS(B, deg) = \begin{bmatrix} 15.843 \\ 355.601 \\ 124.616 \end{bmatrix}$$

Exhibit 5-23

Coordinate Transformations

From the electronic book on Advanced Mathematics

Spherical to Rectangular: $(r, \phi, \theta) \longrightarrow (x, y, z)$

$$\text{StoR}(R, \text{measure}) := \begin{bmatrix} R_0 \cdot \cos(R_1 \cdot \text{measure}) \cdot \sin(R_2 \cdot \text{measure}) \\ R_0 \cdot \sin(R_1 \cdot \text{measure}) \cdot \sin(R_2 \cdot \text{measure}) \\ R_0 \cdot \cos(R_2 \cdot \text{measure}) \end{bmatrix}$$

Example:

$$C := \begin{bmatrix} 5 \\ \dfrac{3 \cdot \pi}{2} \\ \dfrac{\pi}{3} \end{bmatrix} \qquad \text{StoR}(C, \text{rad}) = \begin{bmatrix} 0 \\ -4.33 \\ 2.5 \end{bmatrix}$$

Or

$$D := \begin{bmatrix} 2 \\ \dfrac{\pi}{6} \\ \dfrac{\pi}{4} \end{bmatrix} \qquad \text{StoR}(D, \text{rad}) = \begin{bmatrix} 1.225 \\ 0.707 \\ 1.414 \end{bmatrix}$$

Exhibit 5-24

The functions RtoC and CtoR can be adapted to convert between two-dimensional rectangular and polar coordinates.

Rectangular: **Polar:**

(x, y)   (r, θ)

Coordinate Transformations

From the electronic book on Advanced Mathematics

Rectangular to Polar: **(x, y) $\longrightarrow$ (r, θ)**

$$\text{RtoP}(X, \text{measure}) := \begin{bmatrix} |X| \\ \text{if}\left(|X|, \dfrac{\text{angle}(X_0, X_1)}{\text{measure}}, 0\right) \end{bmatrix}$$

Example:

$$E := \begin{bmatrix} 3 \\ 4 \end{bmatrix} \qquad \text{RtoP}(E, \deg) = \begin{bmatrix} 5 \\ 53.13 \end{bmatrix}$$

Or

$$F := \begin{bmatrix} .1 \\ 3 \end{bmatrix} \qquad \text{RtoP}(F, \deg) = \begin{bmatrix} 3.002 \\ 88.091 \end{bmatrix}$$

Exhibit 5-25

Coordinate Transformations

From the electronic book on Advanced Mathematics

Polar to Rectangular $(r, \theta) \longrightarrow (x, y)$

$$\text{PtoR}(R, \text{measure}) := \begin{bmatrix} R_0 \cdot \cos\left(R_1 \cdot \text{measure}\right) \\ R_0 \cdot \sin\left(R_1 \cdot \text{measure}\right) \end{bmatrix}$$

Example:

$$F := \begin{bmatrix} 2 \\ \dfrac{\pi}{3} \end{bmatrix} \qquad \text{PtoR}(F, \text{rad}) = \begin{bmatrix} 1 \\ 1.732 \end{bmatrix}$$

Or

$$G := \begin{bmatrix} 2 \\ \dfrac{\pi}{2} \end{bmatrix} \qquad \text{PtoR}(G, \text{rad}) = \begin{bmatrix} 0 \\ 2 \end{bmatrix}$$

Exhibit 5-26

Vectors and Matrices and Some Applications to Finance, Engineering and Science

Introduction

In this chapter we dig into an especially rich vein of mathematics which the power of Mathcad can help to mine. The primary aim here is to gain an acquaintance with Mathcad's toolkit of methods for your work with vectors and matrices. Vectors and matrices (or *arrays,* as we will also call them) have many practical uses in commerce and finance, engineering, science, and mathematics, and Mathcad has a particularly strong set of features to help you work with them. It will not be possible to show every feature or describe every possible use, but the tools you are most likely to need will be introduced and examples of how they might help with your own work will be given. Additional information about working with arrays in Mathcad is given in the Mathcad Treasury. You will probably want to continue your investigation of Mathcad's vector and matrix features by exploring the Treasury after finishing this chapter.

Creating a Vector or a Matrix

There are several ways to create a vector or matrix in Mathcad. The straightforward method is to use the "Insert" menu and select "Matrix." Mathcad then presents you with a dialog box that allows you to select the number of rows or columns. Once you have chosen the number of rows and columns and have clicked on "OK," Mathcad inserts a template with placeholders for the size of matrix you have selected. The maximum size for a matrix created in this way is 100 elements—so a 10×10 would be the largest square matrix you can create this way. You can create and work with larger matrices, but you will have to use one of the other techniques for defining and filling shown later in this chapter. In any event, the tedium of typing matrices of more than 100 elements into your Mathcad documents would probably make 100 elements a practical maximum even if Mathcad had not set it as the limit. The maximum possible number of elements in an array in Mathcad is 8 million. For most systems, though, the effective maximum will be somewhere between 1 million and 8 million elements—

A 3 x 5 matrix

$$\begin{bmatrix} 1 & 4 & 7 & 10 & 13 \\ 2 & 5 & 8 & 11 & 14 \\ 3 & 6 & 9 & 12 & 15 \end{bmatrix}$$

A 5 x 3 matrix

$$\begin{bmatrix} 1 & 6 & 11 \\ 2 & 7 & 12 \\ 3 & 8 & 13 \\ 4 & 9 & 14 \\ 5 & 10 & 15 \end{bmatrix}$$

A 5 element vector

$$\begin{bmatrix} 15 \\ 25 \\ 35 \\ 45 \\ 55 \end{bmatrix}$$

Exhibit 6-1

with available system memory being the primary constraint. In systems with limited memory, and when you define several large matrices, the limit could be lower than 1 million elements. If you try to define a matrix that is too large for your system, you will get a "not enough memory" message. Within the available limit, the matrix elements can be distributed in any combination of columns and rows.

Terminology

You will recall that a matrix is named by its number of rows and columns. The first number is the number of rows, so a 3×5 matrix has 3 rows and 5 columns (see Exhibit 6-1). A matrix with the same number of rows and columns is called a *square* matrix (see Exhibit 6-2). Each entry (or element) in a matrix is called by the name of the matrix with subscripts describing the row and column in which that element appears.

A word about nomenclature is in order here. In some linear algebra books, a horizontal row of numbers is called a *row vector* and a vertical list of numbers is called a *column vector.* In Mathcad, only vertical columns of numbers are considered to be vectors. Horizontal rows of numbers are not called *vectors* in Mathcad. It is not likely to make much practical difference, but you should be aware of the semantics, just to avoid confusion.

Another thing to be aware of is the way Mathcad references elements of vectors and matrices. The first row (the one on top) of a vector or matrix is considered to be row 0, followed by row 1, row 2, and so on. Similarly, the left-hand column is column 0, followed by column 1, column 2, and so forth. So the element of an array a which is located at the top left corner is element $a_{0,0}$ rather than $a_{1,1}$ as you might expect. It is possible to change this, if you prefer to have a different referencing scheme. We will discuss how to do this later under the heading of how to change the origin.

Methods for Working with Vectors and Matrices

Ways to create and fill arrays

For many practical problems you might encounter, arrays limited to 100 elements are too small. So Mathcad provides alternative ways of defining and entering data in array.

A square matrix (4 x 4)

$$M := \begin{bmatrix} 12.1 & 52 & 11 & 15 \\ 21 & 63.75 & 12 & 16.3 \\ 33.5 & 76 & 13 & 17 \\ 45 & 88 & 14.8 & 18 \end{bmatrix}$$

Elements of the above matrix

$M_{1,1} = 63.75$

$M_{3,3} = 18$

Exhibit 6-2

$a_{50,3} := 5$

$$a =$$

	0	1	2	3
36	0	0	0	0
37	0	0	0	0
38	0	0	0	0
39	0	0	0	0
40	0	0	0	0
41	0	0	0	0
42	0	0	0	0
43	0	0	0	0
44	0	0	0	0
45	0	0	0	0
46	0	0	0	0
47	0	0	0	0
48	0	0	0	0
49	0	0	0	0
50	0	0	0	5

$a_{2,38} := 4$

$$a =$$

	0	1	2	3	4	5	6	7	8	9	10	11	12	13
0	0	0	0	0	0	0	0	0	0	0	0	0	0	0
1	0	0	0	0	0	0	0	0	0	0	0	0	0	0
2	0	0	0	0	0	0	0	0	0	0	0	0	0	0
3	0	0	0	0	0	0	0	0	0	0	0	0	0	0
4	0	0	0	0	0	0	0	0	0	0	0	0	0	0
5	0	0	0	0	0	0	0	0	0	0	0	0	0	0
6	0	0	0	0	0	0	0	0	0	0	0	0	0	0
7	0	0	0	0	0	0	0	0	0	0	0	0	0	0
8	0	0	0	0	0	0	0	0	0	0	0	0	0	0
9	0	0	0	0	0	0	0	0	0	0	0	0	0	0
10	0	0	0	0	0	0	0	0	0	0	0	0	0	0
11	0	0	0	0	0	0	0	0	0	0	0	0	0	0
12	0	0	0	0	0	0	0	0	0	0	0	0	0	0
13	0	0	0	0	0	0	0	0	0	0	0	0	0	0
14	0	0	0	0	0	0	0	0	0	0	0	0	0	0

Exhibit 6-3

One way to create an array is to use assignment statements to define values for some elements. Once some elements have been defined, Mathcad creates the smallest array that will accommodate the newly defined elements. In Exhibit 6-3 we see how this is done. We first assigned a value of 5 to element $a_{50,3}$ of a matrix we have not yet created. Mathcad recognized that no such matrix existed, and it created the smallest vector that could encompass the element or elements to which we have assigned values. You can see what Mathcad has created by evaluating the array name (typing the name followed by an equal sign). In this example, we went on to define another element ($a_{2,38}$) which caused Mathcad to expand (add more columns to) the array it had previously produced. We could now create assignment statements to give values to other elements of the matrix. For large matrices, this method is most useful if the array contains a lot of elements that are zero.

A second useful way to create and populate arrays is a technique reminiscent of mainframe computer days, but one that is very valuable for the types of large matrices often encountered in real-life problems. For this method, you start by creating an assignment statement in a Mathcad document. The assignment statement tells Mathcad to read from a certain file, and create a matrix with the data from the file. In our example in Exhibit 6-4, the statement is "READPRN(data1)." Mathcad will now take data from a file called "data1" (the exact name is "data1.prn") and create an array named "C" with that data.

To use this method, you must first have created a data file of a specific kind and put it in the directory where Mathcad will expect it to be. The file can be rather freeform. A file of numbers produced by a word processor will do if arranged in rows (the columns needn't line up). But the file must be located in the same directory (or *folder,* as it's called in Windows 95), as your Mathcad program file. Also, the name must have the three-letter extension "prn." If you do not follow this requirement precisely, Mathcad will be unable to locate and use the file and you will see an error message. (*Tip:* To create the data1.prn file, the authors used a word processor and then saved it as a DOS file. But that produced a file that was called "data1.txt." So we went to a DOS window and edited the data1.txt file-changing the name to data1.prn and resaved it. We did not change the contents of the file, only the name. We then copied the file to the Winmcad directory, where the Mathcad program resides. Mathcad now reads the file and uses its data to create an array with the name you have assigned. The "data1.prn" file is shown in Exhibit 6-5.)

A third way to fill the array is to use range variables to assign values to the elements. This is useful for files in which there is a mathematical expression that can be used to generate the values. It is also useful for vectors where all elements are to have the same value (1 for example), or the elements have values that are consecutive integers (see Exhibits 6-6 to 6-8).

Adding and deleting rows or columns

Once you have created a matrix, you may discover that you need to expand it by adding rows and columns. Mathcad has an easy way to do this. After making a copy of your matrix in case

---

C := READPRN(data1)

Exhibit 6-4

```
1   2  3  4   5  6  7  8  9  10 11 12 13 14 15 16 17 18 19 20
2   4  7  8  14 23 25 17 12 13 15 16  2  7 32 12 14 11 19 20
3   2  3  4   5 31  7  8  9 10 11 17 13 14 15 16 17 16 19 20
2   5  7 33  14 13 25 17 12 27 15 16  2  7 11 11 14 11 19 20
1   2  3  4   3  6  7  8  7 10 11 12  1 14  0 16 17 17 19 20
2   5  7  8   0 13 25 17 12  1 15 16  2  7 11 12 14 11 19 20
1   2  3  4   5  6  7  8  3 10 11 12 13 14  1 16 17 18 19 20
7   5  7  8   1 13 25 17 12  1 15 16  2  3 11 12 14 11 19 20
1   2  3  4   5  6  7  8  9  8 11 12 13 14 15 16  7 18 19 20
2   5  7  8  11 13 25 17 12 13 15 16  5  7 11 12 14 11 19 20
1   2  2  4   5  6  7  8  5 10 11 12 13 14 15 19 17 18 19 20
2   5  7  3  14 13 25 17 12  0 15 16  2  7 11 12  0 11 19 20
1   2  3  3   5  6  7  8  7 10 11 12 13 14  5 16 17 18 19 20
2   5  7  8  14 11 25 17 12 13  0 16  2  7 11  1 14 11 19 20
1   2  3  0   5  6  7  8  9 10 11 11 13 14 15 18 17 18 19 20
2   5  7 27  14 13 25 17 12 13 15 16  1  7 11 12 14 11 19 20
1   5  3  4   5  6  7  8  9 10 11  1 13 14 15 16 17 18 19 20
13  5  7  8  14 13  2 17 12 13 15 16  2  7 22 12 14 11 19 20
0   2  3  4   5  6  7  8  9 10 11 12  7 14 15 16 17 18 19 20
9   5  7  8  14 13 25 17 12 13 23 16  2  7 11 12 14 11 19 20
```

Exhibit 6-5

$a := 0 .. 9$

$A_a := 1$

$$A = \begin{array}{c|c} & 0 \\ \hline 0 & 1 \\ 1 & 1 \\ 2 & 1 \\ 3 & 1 \\ 4 & 1 \\ 5 & 1 \\ 6 & 1 \\ 7 & 1 \\ 8 & 1 \\ 9 & 1 \end{array}$$

Exhibit 6-6

$a := 0 .. 9$

$A_a := a$

$$A = \begin{array}{c|c} & 0 \\ \hline 0 & 0 \\ 1 & 1 \\ 2 & 2 \\ 3 & 3 \\ 4 & 4 \\ 5 & 5 \\ 6 & 6 \\ 7 & 7 \\ 8 & 8 \\ 9 & 9 \end{array}$$

Exhibit 6-7

$a := 0 .. 7$

$V_a := a$

$$V = \begin{bmatrix} 0 \\ 1 \\ 2 \\ 3 \\ 4 \\ 5 \\ 6 \\ 7 \end{bmatrix}$$

Or

$V1_a := 1$

$$V1 = \begin{bmatrix} 1 \\ 1 \\ 1 \\ 1 \\ 1 \\ 1 \\ 1 \\ 1 \end{bmatrix}$$

Exhibit 6-8

To add rows and columns to a square matrix (4 x 4)
Select an element of the matrix, then use the Insert menu

$$M := \begin{bmatrix} 12.1 & 52 & 11 & 0 & 0 & 15 \\ 21 & 63.75 & 12 & 0 & 0 & 16.3 \\ 33.5 & 76 & 13 & 0 & 0 & 17 \\ 0 & 0 & 0 & 0 & 0 & 0 \\ 0 & 0 & 0 & 0 & 0 & 0 \\ 0 & 0 & 0 & 0 & 0 & 0 \\ 45 & 88 & 14.8 & 0 & 0 & 18 \end{bmatrix}$$

Exhibit 6-9

you make a mistake in enlarging it, you should select an element of the matrix and call up the "Matrix" selection on the Insert menu. Then indicate the number of rows and columns you want to add, and click on "Insert" (see Exhibit 6-9). The number of rows you have specified will be added below the matrix element you previously selected—and the number of columns you have specified will be added to the right of the matrix element you selected.

If you need to reduce the number of rows and columns, use "Delete." Try it on one of your matrices, but make a copy first if it is an important matrix, in case you are not able to use "Undo" to reverse the changes you made with Delete. Select an element of the matrix, then use "Delete" from the Matrix choice on the Insert menu, after you select the number of rows and columns to be deleted. The rows to be deleted will be the row in which the selected element is located, plus enough rows below the selected element to satisfy your request. The columns to be deleted will start with the column in which the selected element is located and continue to the right until the number of columns requested have been deleted (see Exhibit 6-10).

Other things you can do with matrices

The next two functions enable you to take two existing matrices and join them into a new, larger matrix.

The augment function is illustrated in Exhibit 6-11. This function allows you to take two existing arrays and join them side by side into a new, larger array. The arguments for the augment function are the names of the two matrices to be joined (or the matrices themselves) separated by a comma. The two arrays to be joined must have the same number of rows.

The stack function is illustrated in Exhibit 6-12. It is very similar to the augment function in that it joins two existing arrays into a new, larger array. The difference is that the first matrix becomes the top part of the new matrix, while the second matrix becomes the bottom of the new matrix. This is unlike the augment function, which joined matrices side by side. The arrays to be joined must have the same number of columns.

The two previous operators allow you to join matrices into larger matrices. Our next operator, submatrix, enables you to take subsets out of matrices, thus creating new, smaller matrices. To use submatrix (Exhibit 6-13), you provide arguments that specify

---

To delete 1 row and 1 column from the previous matrix select an element of the matrix, (here the number 13 on the previous matrix) then use the Insert menu to delete 1 row and 1 column.

$$
M := \begin{bmatrix}
12.1 & 52 & 0 & 0 & 15 \\
21 & 63.75 & 0 & 0 & 16.3 \\
0 & 0 & 0 & 0 & 0 \\
0 & 0 & 0 & 0 & 0 \\
0 & 0 & 0 & 0 & 0 \\
45 & 88 & 0 & 0 & 18
\end{bmatrix}
$$

Exhibit 6-10

Augment

$$M := \begin{bmatrix} 12.1 & 52 & 11 & 15 \\ 21 & 63.75 & 12 & 16.3 \\ 33.5 & 76 & 13 & 17 \\ 45 & 88 & 14.8 & 18 \end{bmatrix}$$

$$M1 := \begin{bmatrix} 1 & 5 & 9 & 13 \\ 2 & 6 & 10 & 14 \\ 3 & 7 & 11 & 15 \\ 4 & 8 & 12 & 16 \end{bmatrix}$$

To join these with the Augment

$$\text{augment}\left[\begin{bmatrix} 12.1 & 52 & 11 & 15 \\ 21 & 63.75 & 12 & 16.3 \\ 33.5 & 76 & 13 & 17 \\ 45 & 88 & 14.8 & 18 \end{bmatrix}, \begin{bmatrix} 1 & 5 & 9 & 13 \\ 2 & 6 & 10 & 14 \\ 3 & 7 & 11 & 15 \\ 4 & 8 & 12 & 16 \end{bmatrix}\right] = \begin{bmatrix} 12.1 & 52 & 11 & 15 & 1 & 5 & 9 & 13 \\ 21 & 63.75 & 12 & 16.3 & 2 & 6 & 10 & 14 \\ 33.5 & 76 & 13 & 17 & 3 & 7 & 11 & 15 \\ 45 & 88 & 14.8 & 18 & 4 & 8 & 12 & 16 \end{bmatrix}$$

Or

$$\text{augment}(M, M1) = \begin{bmatrix} 12.1 & 52 & 11 & 15 & 1 & 5 & 9 & 13 \\ 21 & 63.75 & 12 & 16.3 & 2 & 6 & 10 & 14 \\ 33.5 & 76 & 13 & 17 & 3 & 7 & 11 & 15 \\ 45 & 88 & 14.8 & 18 & 4 & 8 & 12 & 16 \end{bmatrix}$$

Exhibit 6-11

- The matrix from which the new submatrix is to be drawn
- The top and bottom rows that bracket the submatrix
- The left and right columns that bracket the submatrix

Commas are used as *delimiters*—to separate the arguments. Remember that the topmost row is row 0 and the leftmost column is column 0.

Another operator that takes elements from an array is the column operator. This operator is invoked using a symbol on the "Vector and Matrix" palette. In Exhibits 6-14 and 6-15, we see an example of the column operator. After clicking on the symbol for the column operator

Stack

$$M := \begin{bmatrix} 12.1 & 52 & 11 & 15 \\ 21 & 63.75 & 12 & 16.3 \\ 33.5 & 76 & 13 & 17 \\ 45 & 88 & 14.8 & 18 \end{bmatrix}$$

$$M1 := \begin{bmatrix} 1 & 5 & 9 & 13 \\ 2 & 6 & 10 & 14 \\ 3 & 7 & 11 & 15 \\ 4 & 8 & 12 & 16 \end{bmatrix}$$

To join these with the Stack

$$\text{stack}\left(\begin{bmatrix} 12.1 & 52 & 11 & 15 \\ 21 & 63.75 & 12 & 16.3 \\ 33.5 & 76 & 13 & 17 \\ 45 & 88 & 14.8 & 18 \end{bmatrix}, \begin{bmatrix} 1 & 5 & 9 & 13 \\ 2 & 6 & 10 & 14 \\ 3 & 7 & 11 & 15 \\ 4 & 8 & 12 & 16 \end{bmatrix} \right) = \begin{bmatrix} 12.1 & 52 & 11 & 15 \\ 21 & 63.75 & 12 & 16.3 \\ 33.5 & 76 & 13 & 17 \\ 45 & 88 & 14.8 & 18 \\ 1 & 5 & 9 & 13 \\ 2 & 6 & 10 & 14 \\ 3 & 7 & 11 & 15 \\ 4 & 8 & 12 & 16 \end{bmatrix}$$

$$\text{stack}(M, M1) = \begin{bmatrix} 12.1 & 52 & 11 & 15 \\ 21 & 63.75 & 12 & 16.3 \\ 33.5 & 76 & 13 & 17 \\ 45 & 88 & 14.8 & 18 \\ 1 & 5 & 9 & 13 \\ 2 & 6 & 10 & 14 \\ 3 & 7 & 11 & 15 \\ 4 & 8 & 12 & 16 \end{bmatrix}$$

Exhibit 6-12

$$A := \begin{bmatrix} 1 & 8 & 15 & 22 & 28 & 36 & 43 & 50 \\ 2 & 9 & 16 & 23 & 30 & 37 & 44 & 51 \\ 3 & 10 & 17 & 24 & 31 & 38 & 45 & 52 \\ 4 & 11 & 18 & 25 & 32 & 39 & 46 & 53 \\ 5 & 12 & 19 & 26 & 33 & 40 & 47 & 54 \\ 6 & 13 & 20 & 27 & 34 & 41 & 48 & 55 \\ 7 & 14 & 21 & 28 & 35 & 42 & 49 & 56 \end{bmatrix}$$

$$B := \text{submatrix}(A, 2, 5, 3, 5) \qquad B = \begin{bmatrix} 24 & 31 & 38 \\ 25 & 32 & 39 \\ 26 & 33 & 40 \\ 27 & 34 & 41 \end{bmatrix}$$

$$C := \text{submatrix}(A, 0, 2, 4, 6) \qquad C = \begin{bmatrix} 28 & 36 & 43 \\ 30 & 37 & 44 \\ 31 & 38 & 45 \end{bmatrix}$$

Exhibit 6-13

(called "Matrix Column" on the palette) and entering the name of the matrix and the number of the column to be operated on, the column operator will create a vector equal to the specified column of the original matrix. You may be wondering if there is a corresponding operator for rows. (*Hint:* Look at the section on the transpose operator, later in this chapter.)

You may wish to change the way Mathcad labels rows and columns of arrays. Mathcad

$$A := \begin{bmatrix} 1 & 4 & 7 \\ 2 & 5 & 8 \\ 3 & 6 & 9 \end{bmatrix}$$

$$A^{<2>} = \begin{bmatrix} 7 \\ 8 \\ 9 \end{bmatrix}$$

Exhibit 6-14

$$A := \begin{bmatrix} 1 & 4 & 7 \\ 2 & 5 & 8 \\ 3 & 6 & 9 \end{bmatrix}$$

$$A^{<1>} = \begin{bmatrix} 4 \\ 5 \\ 6 \end{bmatrix}$$

$$A = \begin{bmatrix} 1 & 4 & 7 \\ 2 & 5 & 8 \\ 3 & 6 & 9 \end{bmatrix}$$

ORIGIN := 1

$$B := A^{<2>}$$

$$M := \begin{bmatrix} 1 & 4 & 7 \\ 2 & 5 & 8 \\ 3 & 6 & 9 \end{bmatrix}$$

$$B = \begin{bmatrix} 7 \\ 8 \\ 9 \end{bmatrix}$$

$$M_{1,1} = 1$$

Exhibit 6-15   **Exhibit 6-16**

---

allows you to do this by a process called "changing the origin." For example, you may want the count of rows and of columns to start with 1 (or some other number) instead of Mathcad's default setting of 0. To do this, type "ORIGIN:= 1" or whatever other value you want it to equal. Mathcad will now call the top row, "row 1" and the top column, "column 1." See Exhibit 6-16 for an example of how this works.

Operations with vectors and matrix operations

Being able to perform complex mathematical operations on arrays of numbers by issuing simple commands is the key to the power of arrays in both mathematics and Mathcad. Mathcad makes that theoretical power available to you in actual calculations. Mathcad makes many operations with vectors and matrices almost as easy as the corresponding arithmetic operations. In addition, the nature of arrays allows a number of mathematical operations that do not apply to ordinary numbers, thereby accounting for some of the richness of mathematics with arrays.

Vector and matrix arithmetic

Adding arrays with Mathcad (see Exhibits 6-17 and 6-18) is as simple as adding numbers if the arrays are the same shape (i.e., have the same number of rows and the same number of

$$
\begin{bmatrix} 1 & 4 & 7 \\ 2 & 5 & 8 \\ 3 & 6 & 9 \end{bmatrix} + 1 =
\begin{bmatrix} 2 & 5 & 8 \\ 3 & 6 & 9 \\ 4 & 7 & 10 \end{bmatrix}
$$

$$
\begin{bmatrix} 1 & 4 & 7 \\ 2 & 5 & 8 \\ 3 & 6 & 9 \end{bmatrix} +
\begin{bmatrix} 1 & 4 & 7 \\ 2 & 5 & 8 \\ 3 & 6 & 9 \end{bmatrix} =
\begin{bmatrix} 2 & 8 & 14 \\ 4 & 10 & 16 \\ 6 & 12 & 18 \end{bmatrix}
$$

$$
\begin{bmatrix} 1 & 4 & 7 \\ 2 & 5 & 8 \\ 3 & 6 & 9 \end{bmatrix} +
\begin{bmatrix} 1 & 4 & 7 \\ 2 & 5 & 8 \\ 3 & 6 & 9 \end{bmatrix} +
\begin{bmatrix} 1 & 4 & 7 \\ 2 & 5 & 8 \\ 3 & 6 & 9 \end{bmatrix} =
\begin{bmatrix} 3 & 12 & 21 \\ 6 & 15 & 24 \\ 9 & 18 & 27 \end{bmatrix}
$$

Symbolically

$$
\begin{bmatrix} 1 & x & 7 \\ 2 & 5 & 8 \\ 3 & 6 & 9+y \end{bmatrix} +
\begin{bmatrix} 1 & 4 & 7 \\ 2 & 5 & 8 \\ 3 & 6 & 9 \end{bmatrix}
$$

$$
\begin{bmatrix} 2 & x+4 & 14 \\ 4 & 10 & 16 \\ 6 & 12 & 18+y \end{bmatrix}
$$

And adding a scalar
(numerical)

$$
\begin{bmatrix} 1 & 4 & 7 \\ 2 & 5 & 8 \\ 3 & 6 & 9 \end{bmatrix} + 1 =
\begin{bmatrix} 2 & 5 & 8 \\ 3 & 6 & 9 \\ 4 & 7 & 10 \end{bmatrix}
$$

Exhibit 6-17

$$A := \begin{bmatrix} 1 & 12 & 101 \\ 11 & 34 & 65 \\ 22 & 5 & 39 \end{bmatrix}$$

$$B := \begin{bmatrix} 1 & 2 & 3 \\ 4 & 5 & 6 \\ 3 & 4 & 7 \end{bmatrix}$$

Complex numbers allowed

$$M := \begin{bmatrix} 1 & 4 & 7 \\ 2 & 5 & 8 \\ 3 & 6 & 9 \end{bmatrix}$$

$$A - B = \begin{bmatrix} 0 & 10 & 98 \\ 7 & 29 & 59 \\ 19 & 1 & 32 \end{bmatrix}$$

$$M1 := \begin{bmatrix} 0 & 0 & 0 \\ i & 2{\cdot}i & 3{\cdot}i \\ 0 & 0 & 0 \end{bmatrix}$$

$$10 - A = \begin{bmatrix} 9 & -2 & -91 \\ -1 & -24 & -55 \\ -12 & 5 & -29 \end{bmatrix}$$

$$M + M1 = \begin{bmatrix} 1 & 4 & 7 \\ 2 + 1i & 5 + 2i & 8 + 3i \\ 3 & 6 & 9 \end{bmatrix}$$

$$A - 10 = \begin{bmatrix} -9 & 2 & 91 \\ 1 & 24 & 55 \\ 12 & -5 & 29 \end{bmatrix}$$

Exhibit 6-18   **Exhibit 6-19**

---

columns). To add, type the names of the matrices separated by a plus sign "+," just as you would do with ordinary numbers. Mathcad instantly does the work of adding the corresponding elements of the matrices, even if they are very large matrices. Symbolic addition of arrays is allowed in Mathcad. Adding a scalar to an array is also allowed as long as you are using the numerical processor. The resulting matrix has the scalar added to each element of the original.

Complex numbers are allowed in vectors and matrices and can participate in the standard operations on arrays (see Exhibit 6-18) as well as some operations (e.g., complex conjugate) that are specific to complex arrays.

Subtraction involving arrays also requires that the arrays be of the same size (see Exhibit 6-19). Array subtraction is essentially the same as array addition, with the exception of the signs.

Multiplication of two arrays requires that the number of columns in the first matrix equal the number of rows in the second matrix (see Exhibit 6-20). If you remember the definition of matrix multiplication, you will understand the reason for this rule—if not, just accept it as one of the rules of matrix manipulation. Mathcad will warn you if you try to violate it. One important thing to keep in mind is that, unlike the case in regular arithmetic, the product of matrix A times matrix B may not be equal to the product of matrix B times matrix A. In fact, the product of matrix B times matrix A may not exist. You probably recall that matrix multiplication involves a fairly complex series of multiplications and additions for each element

$$A := \begin{bmatrix} 1 & 2 & 3 \\ 4 & 5 & 6 \\ 7 & 8 & 9 \end{bmatrix} \qquad\qquad B := \begin{bmatrix} 6 & 7 & 8 \\ 9 & 10 & 11 \\ 12 & 13 & 14 \end{bmatrix}$$

$$A \cdot B = \begin{bmatrix} 60 & 66 & 72 \\ 141 & 156 & 171 \\ 222 & 246 & 270 \end{bmatrix}$$

$$B \cdot A = \begin{bmatrix} 90 & 111 & 132 \\ 126 & 156 & 186 \\ 162 & 201 & 240 \end{bmatrix}$$

$$A \cdot 10 = \begin{bmatrix} 10 & 20 & 30 \\ 40 & 50 & 60 \\ 70 & 80 & 90 \end{bmatrix}$$

Exhibit 6-20

---

of the resultant array. For large arrays, the amount of arithmetic involved can be truly prodigious and the calculations can be very tedious with substantial possibility of error. Here again, Mathcad does the work for you quickly and accurately. Mathcad allows both numerical and symbolic multiplication, and both the numerical and the symbolic calculation engines of Mathcad allow you to multiply an array by a scalar.

Complex conjugates of arrays

In Exhibit 6-21, we see how Mathcad calculates the complex conjugate of a matrix. You can select the name of the matrix and use the quotation mark (") to invoke the complex conjugate. A multiplication involving a complex matrix and its conjugate is seen in Exhibit 6-22.

Identity matrices

A special square array called the *identity matrix* serves a function similar to the number 1 in ordinary multiplication. Just as a number remains unchanged when multiplied by 1, an array multiplied by an identity matrix is unchanged. Mathcad provides an easy way to create an identity matrix. The function "identity" takes a number as argument. The newly created matrix

$$A := \begin{bmatrix} 2 + 3 \cdot i & 1 - i & 5 \cdot i \\ 1 + i & 6 - i & 1 + 3 \cdot i \\ 5 - 6 \cdot i & 3 & 0 \end{bmatrix}$$

$$\overline{A} = \begin{bmatrix} 2 - 3i & 1 + 1i & -5i \\ 1 - 1i & 6 + 1i & 1 - 3i \\ 5 + 6i & 3 & 0 \end{bmatrix} \qquad \text{Complex Conjugate}$$

$$A \cdot \overline{A} = \begin{bmatrix} -17 + 23i & 6 + 15i & 13 - 14i \\ -3 + 13i & 40 + 11i & 8 - 24i \\ -5 - 30i & 29 + 2i & -27 - 34i \end{bmatrix}$$

$$\overline{A} \cdot A = \begin{bmatrix} -17 - 23i & 6 - 15i & 13 + 14i \\ -3 - 13i & 40 - 11i & 8 + 24i \\ -5 + 30i & 29 - 2i & -27 + 34i \end{bmatrix}$$

Or

$$B := \begin{bmatrix} 1 + 2 \cdot i & i \\ 3 & 2 - 3 \cdot i \end{bmatrix}$$

$$\overline{B} = \begin{bmatrix} 1 - 2i & -1i \\ 3 & 2 + 3i \end{bmatrix}$$

$$B \cdot \overline{B} = \begin{bmatrix} 5 + 3i & -1 + 1i \\ 9 - 15i & 13 - 3i \end{bmatrix}$$

$$\overline{B} \cdot B = \begin{bmatrix} 5 - 3i & -1 - 1i \\ 9 + 15i & 13 + 3i \end{bmatrix}$$

Exhibit 6-21

$$\text{identity}(1) = 1$$

$$\text{identity}(2) = \begin{bmatrix} 1 & 0 \\ 0 & 1 \end{bmatrix}$$

$$\text{identity}(3) = \begin{bmatrix} 1 & 0 & 0 \\ 0 & 1 & 0 \\ 0 & 0 & 1 \end{bmatrix}$$

$$\text{identity}(4) = \begin{bmatrix} 1 & 0 & 0 & 0 \\ 0 & 1 & 0 & 0 \\ 0 & 0 & 1 & 0 \\ 0 & 0 & 0 & 1 \end{bmatrix}$$

$$\text{identity}(5) = \begin{bmatrix} 1 & 0 & 0 & 0 & 0 \\ 0 & 1 & 0 & 0 & 0 \\ 0 & 0 & 1 & 0 & 0 \\ 0 & 0 & 0 & 1 & 0 \\ 0 & 0 & 0 & 0 & 1 \end{bmatrix}$$

$$A := \text{identity}(3) \cdot \begin{bmatrix} 1 & 4 & 7 \\ 2 & 5 & 8 \\ 3 & 6 & 9 \end{bmatrix}$$

$$A = \begin{bmatrix} 1 & 4 & 7 \\ 2 & 5 & 8 \\ 3 & 6 & 9 \end{bmatrix}$$

$$B := \begin{bmatrix} 1 & 4 & 7 \\ 2 & 5 & 8 \\ 3 & 6 & 9 \end{bmatrix} \cdot \text{identity}(3)$$

$$B = \begin{bmatrix} 1 & 4 & 7 \\ 2 & 5 & 8 \\ 3 & 6 & 9 \end{bmatrix}$$

Exhibit 6-22

identity(2000) =

	0	1	2	3	4	5	6	7	8	9	10	11	12	13
0	1	0	0	0	0	0	0	0	0	0	0	0	0	0
1	0	1	0	0	0	0	0	0	0	0	0	0	0	0
2	0	0	1	0	0	0	0	0	0	0	0	0	0	0
3	0	0	0	1	0	0	0	0	0	0	0	0	0	0
4	0	0	0	0	1	0	0	0	0	0	0	0	0	0
5	0	0	0	0	0	1	0	0	0	0	0	0	0	0
6	0	0	0	0	0	0	1	0	0	0	0	0	0	0
7	0	0	0	0	0	0	0	1	0	0	0	0	0	0
8	0	0	0	0	0	0	0	0	1	0	0	0	0	0
9	0	0	0	0	0	0	0	0	0	1	0	0	0	0
10	0	0	0	0	0	0	0	0	0	0	1	0	0	0
11	0	0	0	0	0	0	0	0	0	0	0	1	0	0
12	0	0	0	0	0	0	0	0	0	0	0	0	1	0
13	0	0	0	0	0	0	0	0	0	0	0	0	0	1
14	0	0	0	0	0	0	0	0	0	0	0	0	0	0

Exhibit 6-23

$$A := \begin{bmatrix} 1 & 1 & 1 \\ 0 & 1 & 1 \\ 0 & 0 & 1 \end{bmatrix}$$

$$A^{-1} = \begin{bmatrix} 1 & -1 & 0 \\ 0 & 1 & -1 \\ 0 & 0 & 1 \end{bmatrix}$$

Exhibit 6-24

---

will have its elements along the diagonal, each equal to 1. The identity function is not limited to a 10 × 10 matrix, as you can see from Exhibit 6-23. The 2000 × 2000 identity matrix was approximately the largest the authors could produce on their computers before Mathcad gave an error message. It took several minutes to create this matrix on a 166-MHz Pentium machine. You will notice that Mathcad produced a scrolling display of the output matrix. This is the default for display of large outputs.

The authors multiplied some relatively large identity matrices to investigate the massive computational capability that Mathcad puts at your disposal with simple matrix commands. As you probably know by now, Mathcad (on a relatively fast computer) does most calculations almost instantaneously. This is true even if they involve very complex mathematics, but for matrix multiplication and some other matrix operations, the number of calculations to do one operation can be very large and can grow rapidly with the size of the matrices involved. We found that multiplying two 100 × 100 identity matrices was almost instantaneous, but a 400 × 400 multiplication took a few seconds. Multiplying identity matrices larger than 400 × 400 took many minutes—an indication of the vast amount of computation unleashed by a simple command to multiply two identity matrices. Not surprisingly, the answers to all of these multiplications of identity matrices were new identity matrices of the same size.

Inverses

Calculating the inverse of a matrix is a relatively complex process for even a small matrix and for a large one, can be a very tedious and computationally intensive exercise. Mathcad does the calculation for you effortlessly, if the matrix has an inverse. Simply use the normal Mathcad method of raising a number to a power (in this case the -1 power) by using the caret ("^") to represent exponentiation and then typing in the power. Mathcad also provides, on its "Arithmetic" palette, a symbol for raising to a power. Mathcad will compute the inverse as in Exhibits 6-24 through 6-28. For symbolic calculation of inverses, use the "Matrix" then "Invert" on the Symbolics menu.

Notice that when a matrix is multiplied by its inverse, the result is an identity matrix. Also be aware that (unlike matrix multiplication in general) multiplication by the inverse is allowed in either order and will produce the identity matrix regardless of the order in which the multiplication is done (Exhibit 6-29).

Not all matrices have inverses. Nonsquare matrices do not have inverses. For square matrices, we will see in the section on determinants how to determine if an inverse exists.

Mathcad calculates the inverse of a (square) matrix when you issue the command for raising

$$A := \begin{bmatrix} 1 & .42 & .54 & .66 \\ .42 & 1 & .32 & .44 \\ .54 & .32 & 1 & .22 \\ .66 & .44 & .22 & 1 \end{bmatrix}$$

$$A^{-1} = \begin{bmatrix} 2.5076 & -0.1230 & -1.0115 & -1.3783 \\ -0.1230 & 1.3322 & -0.2614 & -0.4475 \\ -1.0115 & -0.2614 & 1.5318 & 0.4456 \\ -1.3783 & -0.4475 & 0.4456 & 2.0086 \end{bmatrix}$$

$$A \cdot A^{-1} = \begin{bmatrix} 1 & 0 & 0 & 0 \\ 0 & 1 & 0 & 0 \\ 0 & 0 & 1 & 0 \\ 0 & 0 & 0 & 1 \end{bmatrix}$$

$$A^{-1} \cdot A = \begin{bmatrix} 1 & 0 & 0 & 0 \\ 0 & 1 & 0 & 0 \\ 0 & 0 & 1 & 0 \\ 0 & 0 & 0 & 1 \end{bmatrix}$$

Exhibit 6-25

$$A := \begin{bmatrix} 1 + 2 \cdot i & 2 + 3 \cdot i & 4 + .5 \cdot i \\ .5 + .75 \cdot i & 3 + 2.5 \cdot i & 3 + 7 \cdot i \\ 1 + 2 \cdot i & 3 + 2.3 \cdot i & 1 + 5 \cdot i \end{bmatrix}$$

$$A^{-1} = \begin{bmatrix} 0.25 + 0.046i & -0.176 + 0.544i & -6.992 \cdot 10^{-3} - 0.673i \\ -0.14 - 0.262i & 0.147 - 0.248i & 0.093 + 0.369i \\ 0.095 + 0.087i & 0.051 - 0.04i & -0.113 - 0.069i \end{bmatrix}$$

$$A \cdot A^{-1} = \begin{bmatrix} 1 & 0 & 0 \\ 0 & 1 & 0 \\ 0 & 0 & 1 \end{bmatrix}$$

$$B := \begin{bmatrix} 4 & 0 & 5 \\ 0 & 1 & -6 \\ 3 & 0 & 4 \end{bmatrix}$$

$$B^{-1} = \begin{bmatrix} 4 & 0 & -5 \\ -18 & 1 & 24 \\ -3 & 0 & 4 \end{bmatrix}$$

$$B \cdot B^{-1} = \begin{bmatrix} 1 & 0 & 0 \\ 0 & 1 & 0 \\ 0 & 0 & 1 \end{bmatrix}$$

$$B^{-1} \cdot B = \begin{bmatrix} 1 & 0 & 0 \\ 0 & 1 & 0 \\ 0 & 0 & 1 \end{bmatrix}$$

Exhibit 6-26

$$A := \begin{bmatrix}
10 & .555 & -.7 & 2. & 2 & 5.7 & -1 & -5.8 & -1.7 & 8.2 \\
2.4 & 7.1 & -7.1 & 1.6 & .008 & 0 & .8 & 0 & -.47 & -.78 \\
7.55 & 2.98 & -1.7 & -10 & 0 & .89 & -.78 & -.9 & 9 & -.01 \\
1.1 & 2.2 & 0 & 3.3 & 4.4 & 5.5 & 6.6 & 7.7 & 8.8 & 9.9 \\
4.777 & -5.8 & -1 & 100 & 0 & 0 & .777 & 20.8 & -20.8 & 20.8 \\
8.7 & -50.8 & 9.8 & 9.2 & 9.72 & -2 & -1 & 0 & 7.851 & -.7 \\
1.1 & 2.2 & 8.8 & 3.3 & 4.4 & 7.7 & 5.5 & 0 & 0 & 8 \\
15 & 16 & 17 & 18 & 19 & 20 & 21 & 0 & -.01 & -.02 \\
10 & 20 & -.99 & 9.25 & 9.007 & -4.87 & 1.007 & 4.5 & -89 & 0 \\
0 & 0 & 1.001 & -.1 & -.7 & -.1 & -.52 & -.75 & 75.9 & -1.88
\end{bmatrix}$$

$A^{-1} =$

	0	1	2	3	4	5	6	7	8	9
0	-0.14	0.27	0.24	-0.14	0.02	0.04	0.29	-0.04	0	-0.02
1	0.02	-0.05	-0.02	0.01	0	-0.02	-0.04	0.01	0.02	0.02
2	-0.09	0.03	0.11	-0.09	0.01	0.01	0.18	-0.02	0.01	0.01
3	0	0.02	-0.02	-0.02	0.01	0	0.01	0	0	0.01
4	0.37	-0.63	-0.43	0.33	-0.05	-0.06	-0.71	0.11	0.06	0.09
5	0.56	-1.17	-0.43	0.48	-0.03	-0.16	-1.23	0.22	-0.05	-0.05
6	-0.71	1.5	0.57	-0.58	0.04	0.19	1.49	-0.22	-0.02	-0.04
7	0.2	-0.62	-0.11	0.28	0	-0.08	-0.61	0.1	-0.01	-0.02
8	0	0.01	0	0	0	0	0	0	0	0.01
9	-0.14	0.39	0.12	-0.12	0.01	0.05	0.45	-0.09	0.02	0.02

$A \cdot A^{-1} =$

	0	1	2	3	4	5
0	1	0	0	0	0	0
1	0	1	0	0	0	0
2	0	$1.3 \cdot 10^{-15}$	1	0	0	0
3	0	$-1.42 \cdot 10^{-15}$	0	1	0	0
4	$2.57 \cdot 10^{-15}$	$-5.89 \cdot 10^{-15}$	$-1.25 \cdot 10^{-15}$	$2.41 \cdot 10^{-15}$	1	0
5	0	$1.68 \cdot 10^{-15}$	0	$-1.05 \cdot 10^{-15}$	0	1
6	0	$-1.26 \cdot 10^{-15}$	0	0	0	0
7	$-1.24 \cdot 10^{-15}$	$-4.5 \cdot 10^{-15}$	$-2.4 \cdot 10^{-15}$	$-2.63 \cdot 10^{-15}$	0	0
8	$-2.27 \cdot 10^{-15}$	$-3.57 \cdot 10^{-15}$	$-1.27 \cdot 10^{-15}$	$-2.43 \cdot 10^{-15}$	0	0
9	0	0	0	0	0	0

Exhibit 6-27

Inverse of a complex matrix

$$A := \begin{bmatrix} 4 + 3 \cdot i & 7 - 2 \cdot i \\ 1 + 5 \cdot i & 3 + 8 \cdot i \end{bmatrix}$$

$$A^{-1} = \begin{bmatrix} -0.0254 - 0.2829i & 0.242 - 2.2099 \cdot 10^{-3} \, i \\ -0.0122 + 0.1691i & -0.1017 - 0.1315i \end{bmatrix}$$

Exhibit 6-28

$A := identity(400)$

$A \cdot A =$

	0	1	2	3	4	5	6	7	8	9	10	11	12	13
0	1	0	0	0	0	0	0	0	0	0	0	0	0	0
1	0	1	0	0	0	0	0	0	0	0	0	0	0	0
2	0	0	1	0	0	0	0	0	0	0	0	0	0	0
3	0	0	0	1	0	0	0	0	0	0	0	0	0	0
4	0	0	0	0	1	0	0	0	0	0	0	0	0	0
5	0	0	0	0	0	1	0	0	0	0	0	0	0	0
6	0	0	0	0	0	0	1	0	0	0	0	0	0	0
7	0	0	0	0	0	0	0	1	0	0	0	0	0	0
8	0	0	0	0	0	0	0	0	1	0	0	0	0	0
9	0	0	0	0	0	0	0	0	0	1	0	0	0	0
10	0	0	0	0	0	0	0	0	0	0	1	0	0	0
11	0	0	0	0	0	0	0	0	0	0	0	1	0	0
12	0	0	0	0	0	0	0	0	0	0	0	0	1	0
13	0	0	0	0	0	0	0	0	0	0	0	0	0	1
14	0	0	0	0	0	0	0	0	0	0	0	0	0	0

Exhibit 6-29

it to the power -1, as you saw above—provided an inverse exists. Other integer powers of a matrix (squares, cubes, etc.) are calculated in the same manner.

Division

Division is not well defined for matrices, but division by a scalar is allowed by both the numerical and symbolic engines in Mathcad. In addition, the symbolic engine interprets matrix A divided by matrix B to be equivalent to matrix A times the inverse of Matrix B. See Exhibit 6-30 for examples.

Powers of matrices

Mathcad can raise square matrices to any positive or negative integer power. You have already seen that when you tell Mathcad to compute the -1 power of a matrix, it returns the inverse (provided an inverse exists). By definition, any matrix raised to the 0 power produces an

---

$$A := \begin{bmatrix} 1 & 2 & 3 \\ 4 & 5 & 6 \\ 7 & 8 & 9 \end{bmatrix}$$

$$\frac{A}{2} = \begin{bmatrix} 0.5 & 1 & 1.5 \\ 2 & 2.5 & 3 \\ 3.5 & 4 & 4.5 \end{bmatrix}$$

$$B := \begin{bmatrix} 1 & 1 & 1 \\ 0 & 1 & 1 \\ 0 & 0 & 1 \end{bmatrix}$$

Symbolically

$$\frac{\begin{bmatrix} 1 & 2 & 3 \\ 4 & 5 & 6 \\ 7 & 8 & 9 \end{bmatrix}}{\begin{bmatrix} 1 & 1 & 1 \\ 0 & 1 & 1 \\ 0 & 0 & 1 \end{bmatrix}}$$

$$\begin{bmatrix} 1 & 1 & 1 \\ 4 & 1 & 1 \\ 7 & 1 & 1 \end{bmatrix}$$ Which is equivalent to $$A \cdot B^{-1} = \begin{bmatrix} 1 & 1 & 1 \\ 4 & 1 & 1 \\ 7 & 1 & 1 \end{bmatrix}$$

Exhibit 6-30

identity matrix of the same size. In Exhibit 6-31, you see examples of using Mathcad to raise a matrix to some positive powers. Exhibit 6-32 shows some analogous results for negative powers.

Systems of linear equations

In Exhibits 6-33 to 6-35, we see some examples of how to solve a system of linear equations by creating a matrix (we call it "A") of the coefficients of the unknowns and vector (called "B") of the constants on the right side of the equal sign. To get the solution, we then find the inverse of A and multiply by that inverse times B. The answer is a vector of values for the variables (in this case four of them). To check these results, we substituted the calculated values back into the original equations. A complete discussion of this topic is beyond the scope of this book, but numerous linear algebra books are available to supply the complete theoretical framework.

Powers

$$A := \begin{bmatrix} 1 & 4 & 7 \\ 2 & 5 & 8 \\ 3 & 6 & 9 \end{bmatrix}$$

$$A^2 = \begin{bmatrix} 30 & 66 & 102 \\ 36 & 81 & 126 \\ 42 & 96 & 150 \end{bmatrix}$$

$$A^3 = \begin{bmatrix} 468 & 1.062 \cdot 10^3 & 1.656 \cdot 10^3 \\ 576 & 1.305 \cdot 10^3 & 2.034 \cdot 10^3 \\ 684 & 1.548 \cdot 10^3 & 2.412 \cdot 10^3 \end{bmatrix}$$

$$A^4 = \begin{bmatrix} 7.56 \cdot 10^3 & 1.712 \cdot 10^4 & 2.668 \cdot 10^4 \\ 9.288 \cdot 10^3 & 2.103 \cdot 10^4 & 3.278 \cdot 10^4 \\ 1.102 \cdot 10^4 & 2.495 \cdot 10^4 & 3.888 \cdot 10^4 \end{bmatrix}$$

Exhibit 6-31

$$B := \begin{bmatrix} 1 & 1 & 1 \\ 0 & 1 & 1 \\ 0 & 0 & 1 \end{bmatrix}$$

$$B^{-1} = \begin{bmatrix} 1 & -1 & 0 \\ 0 & 1 & -1 \\ 0 & 0 & 1 \end{bmatrix}$$

$$B^{-2} = \begin{bmatrix} 1 & -2 & 1 \\ 0 & 1 & -2 \\ 0 & 0 & 1 \end{bmatrix}$$

$$B^{-3} = \begin{bmatrix} 1 & -3 & 3 \\ 0 & 1 & -3 \\ 0 & 0 & 1 \end{bmatrix}$$

$$B^{-4} = \begin{bmatrix} 1 & -4 & 6 \\ 0 & 1 & -4 \\ 0 & 0 & 1 \end{bmatrix}$$

Exhibit 6-32

Simultaneous linear equations

$$2 \cdot x_1 + x_2 + 5 \cdot x_3 + x_4 = 5$$
$$x_1 + x_2 - 3 \cdot x_3 - 4 \cdot x_4 = -1$$
$$3 \cdot x_1 + 6 \cdot x_2 - 2 \cdot x_3 + x_4 = 8$$
$$2 \cdot x_1 + 2 \cdot x_2 + 2 \cdot x_3 - 3 \cdot x_4 = 2$$

$$A := \begin{bmatrix} 2 & 1 & 5 & 1 \\ 1 & 1 & -3 & -4 \\ 3 & 6 & -2 & 1 \\ 2 & 2 & 2 & -3 \end{bmatrix} \qquad B := \begin{bmatrix} 5 \\ -1 \\ 8 \\ 2 \end{bmatrix}$$

So the solution is

$$A^{-1} \cdot B = \begin{bmatrix} 2 \\ 0.2 \\ 0 \\ 0.8 \end{bmatrix}$$

To check, we substitute these values for the four variables:

$$2 \cdot 2 + .2 + 5 \cdot 0 + .8 = 5$$

$$2 + .2 + -3 \cdot 0 - 4 \cdot .8 = -1$$

$$3 \cdot 2 + 6 \cdot .2 - 2 \cdot 0 + .8 = 8$$

$$2 \cdot 2 + 2 \cdot .2 + 2 \cdot 0 - 3 \cdot .8 = 2$$

Exhibit 6-33

The transpose

Mathcad's transpose operator, demonstrated in Exhibit 6-36, flips the array so that the rows of the original array become the columns of the new array and the columns of the original become rows. An important relationship that involves the transpose operation ("the transpose of the product of two matrices equals the product of the transposed matrices taken in reverse

Simultaneous Linear Equations

$$A := \begin{bmatrix} 1 & 4 & 3 \\ 2 & 5 & 4 \\ 1 & -3 & -2 \end{bmatrix}$$

$$B := \begin{bmatrix} 1 \\ 4 \\ 5 \end{bmatrix}$$

$$A^{-1} \cdot B = \begin{bmatrix} 3 \\ -2 \\ 2 \end{bmatrix}$$

Exhibit 6-34

order") is also illustrated here using the Mathcad transpose operator. Exhibit 6-37 shows how the column operator, which we saw earlier, can be used in conjunction with "transpose" to extract a row from a matrix.

Determinants

If you know how to calculate determinants for small (square) matrices, you probably remember that calculating determinants gets increasingly complicated and laborious as the size of the matrix increases. In Exhibits 6-38 and 6-39 we see how Mathcad calculates determinants. It

Solve 2 simultaneous linear equations

$$A := \begin{bmatrix} 3.8 & 7.2 \\ 1.3 & -0.9 \end{bmatrix} \qquad B := \begin{bmatrix} 16.5 \\ -22.1 \end{bmatrix}$$

$$A^{-1} \cdot B = \begin{bmatrix} -11.2887 \\ 8.2496 \end{bmatrix}$$

Exhibit 6-35

$$A := \begin{bmatrix} 1 & 4 & 3 \\ 2 & 5 & 4 \\ 1 & -3 & -2 \end{bmatrix}$$

$$A^T = \begin{bmatrix} 1 & 2 & 1 \\ 4 & 5 & -3 \\ 3 & 4 & -2 \end{bmatrix}$$

$$B := \begin{bmatrix} 2 & 5 & 6 \\ 7 & 9 & 4 \\ 3 & 2 & 5 \end{bmatrix}$$

We want to demonstrate an important relationship for matrices:
The transpose of the product of two matrices equals
the product of the transposed matrices taken in reverse order

$$(A \cdot B)^T = \begin{bmatrix} 39 & 51 & -25 \\ 47 & 63 & -26 \\ 37 & 52 & -16 \end{bmatrix}$$

$$B^T \cdot A^T = \begin{bmatrix} 39 & 51 & -25 \\ 47 & 63 & -26 \\ 37 & 52 & -16 \end{bmatrix}$$

Exhibit 6-36

---

uses the symbol from the "Vector and Matrix" palette, which looks like an absolute-value symbol. To avoid confusion, you should be aware that this symbol is one of the most widely employed in Mathcad; it is used for several types of operations, including

- The absolute value of a number
- The absolute value (or norm) of a complex number
- The magnitude of a vector
- The determinant of a square matrix

Simply insert the name of the square matrix in the placeholder, and Mathcad will calculate

$$A := \begin{bmatrix} 1 & 5 & 9 & 13 & 17 & 21 \\ 2 & 6 & 10 & 14 & 18 & 22 \\ 3 & 7 & 11 & 15 & 19 & 23 \\ 4 & 8 & 12 & 16 & 20 & 24 \end{bmatrix}$$

$$\left(A^T\right)^{<2>} = \begin{bmatrix} 3 \\ 7 \\ 11 \\ 15 \\ 19 \\ 23 \end{bmatrix}$$

Exhibit 6-37

the value. An important use of the determinant is to predict whether a square matrix has an inverse. Matrices with determinants equal to zero do not have inverses.

Mathcad functions to use with matrices

Two Mathcad functions return the number of rows and the number of columns for a matrix. If the matrix is called "A," simply type in "rows(A)" to cause Mathcad to return the number of rows or "cols(A)" to have the number of columns returned. See Exhibit 6-40 for an example. For a vector, as you might expect, "cols" will always return a 1. Both "rows" and "cols" will return a 0 if the argument is a scalar. Both "rows" and "cols" can be used with the symbolic engine.

$$A := \begin{bmatrix} 4 & 4 & 8 & 8 \\ 0 & 1 & 2 & 2 \\ 0 & 0 & 2 & 6 \\ 0 & 0 & 0 & 2 \end{bmatrix}$$

Determinant

$$|A| = 16$$

Exhibit 6-38

Determinant is zero (this matrix has no inverse)

$$\left| \begin{bmatrix} 3 & 12 & 21 \\ 6 & 15 & 24 \\ 9 & 18 & 27 \end{bmatrix} \right| = -1.799 \cdot 10^{-14}$$

$$\begin{bmatrix} 3 & 12 & 21 \\ 6 & 15 & 24 \\ 9 & 18 & 27 \end{bmatrix}^{-1} = \begin{bmatrix} 1.501 \cdot 10^{15} & -3.002 \cdot 10^{15} & 1.501 \cdot 10^{15} \\ -3.002 \cdot 10^{15} & 6.005 \cdot 10^{15} & -3.002 \cdot 10^{15} \\ 1.501 \cdot 10^{15} & -3.002 \cdot 10^{15} & 1.501 \cdot 10^{15} \end{bmatrix}$$

Determinant is not zero (this matrix has an inverse)

$$\left| \begin{bmatrix} 1+2 \cdot i & 2+3 \cdot i & 4+.5 \cdot i \\ .5+.75 \cdot i & 3+2.5 \cdot i & 3+7 \cdot i \\ 1+2 \cdot i & 3+2.3 \cdot i & 1+5 \cdot i \end{bmatrix} \right| = -16.7 - 38.463i$$

$$\begin{bmatrix} 1+2 \cdot i & 2+3 \cdot i & 4+.5 \cdot i \\ .5+.75 \cdot i & 3+2.5 \cdot i & 3+7 \cdot i \\ 1+2 \cdot i & 3+2.3 \cdot i & 1+5 \cdot i \end{bmatrix}^{-1} = \begin{bmatrix} 0.25+0.046i & -0.176+0.544i & -6.992 \cdot 10^{-3} - 0.673i \\ -0.14-0.262i & 0.147-0.248i & 0.093+0.369i \\ 0.095+0.087i & 0.051-0.04i & -0.113-0.069i \end{bmatrix}$$

Exhibit 6-39

Given the name of an array, two Mathcad functions ("max" and "min") return, respectively, the largest and the smallest elements of the array as seen in Exhibit 6-41. The "max" function will select the most positive element, not the one with the largest absolute value, and the "min" function will find the most negative element. When a matrix has unknown elements, the symbolic operator eliminates as many candidates as possible for "max" or "min" and returns the remaining possibilities (perhaps including unknowns). For a matrix containing complex numbers, "max" returns a new complex number whose real part is determined by finding the largest real part from among the elements in the array. The imaginary part of the number is the largest imaginary part among the elements in the array. This procedure is necessary because there is no way to say that one complex number is bigger or smaller than another complex number. The "min" function operates in a similar way. It creates a complex number whose real part is the smallest from among the real parts of the elements of the matrix and

$$A := \begin{bmatrix} 1 & 5 & 9 & 13 & 17 & 21 \\ 2 & 6 & 10 & 14 & 18 & 22 \\ 3 & 7 & 11 & 15 & 19 & 23 \\ 4 & 8 & 12 & 16 & 20 & 24 \end{bmatrix}$$

$$A := \begin{bmatrix} 1 & 5 & 9 & 13 & 17 & 21 \\ 2 & 6 & 10 & 14 & 18 & 22 \\ 3 & 7 & -30 & 15 & 19 & 23 \\ 4 & 8 & 12 & 16 & 20 & 24 \end{bmatrix}$$

$\text{rows}(A) = 4$

$\min(A) = -30$

$\text{cols}(A) = 6$

$\max(A) = 24$

Exhibit 6-40

Exhibit 6-41

---

whose imaginary part is the smallest part among the imaginary parts of the elements of the array.

The "tr" function, when given the name of a square array, returns the sum of the elements along the diagonal as in Exhibit 6-42. The "tr" function can also be used in symbolic computation.

Mathcad functions for vectors

Two valuable functions for use with vectors follow:

1. The "last" function returns the subscript of the final entry in the vector. For example, in Exhibit 6-43, "last(v)" returns the answer 7, since the element v_7 is the last element in v.

2. The "length" function returns the number of elements in the vector. In Exhibit 6-44, "length(V)" returns the answer 5, because there are five elements in V.

$$A := \begin{bmatrix} 1 & 8 & 15 & 22 & 28 & 36 & 43 \\ 2 & 9 & 16 & 23 & 30 & 37 & 44 \\ 3 & 10 & 17 & 24 & 31 & 38 & 45 \\ 4 & 11 & 18 & 25 & 32 & 39 & 46 \\ 5 & 12 & 19 & 26 & 33 & 40 & 47 \\ 6 & 13 & 20 & 27 & 34 & 41 & 48 \\ 7 & 14 & 21 & 28 & 35 & 42 & 49 \end{bmatrix}$$

$$v := \begin{bmatrix} 1 \\ 4 \\ 11 \\ 3 \\ 6 \\ 2 \\ 9 \\ 12 \end{bmatrix}$$

$\text{tr}(A) = 175$

Check $\qquad 1 + 9 + 17 + 25 + 33 + 41 + 49 = 175$

$\text{last}(v) = 7$

Exhibit 6-42

Exhibit 6-43

$$V := \begin{bmatrix} 1 \\ 2 \cdot i \\ \pi \\ 72 \\ 5 \end{bmatrix}$$

For a real vector

$$\text{length}(V) = 5$$

$$V := \begin{bmatrix} 2 \\ 5.3 \\ \pi \\ 72 \\ 5 \end{bmatrix}$$

$$\text{length}\left(\begin{bmatrix} 1 \\ 2 \cdot i \\ \pi \\ 72 \\ 5 \end{bmatrix} \right) = 5$$

$$|V| = 72.464$$

Check: $\sqrt{2^2 + 5.3^2 + \pi^2 + 72^2 + 5^2} = 72.464$

Exhibit 6-44   **Exhibit 6-45**

The magnitude (or vector norm) operator again uses the absolute-value symbol. Its argument is a vector, and it returns a number referred to as the *geometric length*. For a vector of real numbers, the geometric length is the square root of the sums of the squares of the elements (see Exhibit 6-45). If the vector has complex elements, the squares of the elements are replaced by the elements times their complex conjugates (or equivalently, the complex norms squared) as in Exhibit 6-46. If the vector has three elements and represents a vector in space (where the ele-

For a vector with 1 or more imaginary or complex elements

$$V := \begin{bmatrix} 1 \\ 2 \cdot i \\ \pi \\ 72 \\ 5 \end{bmatrix}$$

$$|V| = 72.276$$

Check: $\sqrt{1^2 + (|2 \cdot i|)^2 + \pi^2 + 72^2 + 5^2} = 72.276$

Exhibit 6-46

ments are the x, y, and z components), the magnitude represents the physical length of the vector. A little later we will see how the magnitude for such a 3D vector is used to produce a *unit vector*—a vector that points in the same direction as the original vector and has a length of 1.

The dot product operator multiplies two vectors and produces a number (a scalar). There is another type of vector multiplication that has a vector as its result. It is shown in a later section. To multiply two vectors, they must have the same number of elements. The usual multiplication operator "*" can be used to produce a dot product. Mathcad will recognize that if you type "A*B," for example, and both A and B are vectors, you must be asking Mathcad to compute the dot product. You can also order up a dot product by selecting it from the "Vector and Matrix" palette (middle row) and filling in the placeholders. The dot product calculation involves multiplying the two vectors element by element and then summing the results of those multiplications. Exhibit 6-47 illustrates the use of the dot product by showing a vector of ones dotted against a vector of the integers from 1 to 7. In effect, this particular dot product is adding the integers from 1 through 7.

Exhibit 6-48 uses the symbolic processor to compute the dot product of a vector "A" and a vector "B." The main purpose of this example is to show how the dot product is calculated by showing the formula for this seven-element example.

A common use of the dot product for physical vectors is to determine the angle between the vectors—because the dot product of two vectors starting at the same point is equivalent to multiplying the lengths of the vectors times the cosine of the angle between them (see Exhibit 6-49). Clearly, for two perpendicular vectors, the dot product will be zero, since the cosine of the 90° angle between them is 0.

In doing a calculation such as the one just illustrated, you might need to convert degrees and decimals to degrees, minutes, and seconds (Exhibit 6-50).

Dot Product

$$V := \begin{bmatrix} 1 \\ 1 \\ 1 \\ 1 \\ 1 \\ 1 \\ 1 \end{bmatrix} \qquad V1 := \begin{bmatrix} 1 \\ 2 \\ 3 \\ 4 \\ 5 \\ 6 \\ 7 \end{bmatrix}$$

$V \cdot V1 = 28$

Exhibit 6-47

Dot Product

Symbolic processor shows the mathematical expression for computing the dot product for two 7-element vectors

$$\begin{bmatrix} A0 \\ A1 \\ A2 \\ A3 \\ A4 \\ A5 \\ A6 \end{bmatrix} \cdot \begin{bmatrix} B0 \\ B1 \\ B2 \\ B3 \\ B4 \\ B5 \\ B6 \end{bmatrix}$$

$A0 \cdot B0 + A1 \cdot B1 + A2 \cdot B2 + A3 \cdot B3 + A4 \cdot B4 + A5 \cdot B5 + A6 \cdot B6$

Exhibit 6-48

$$u := \begin{bmatrix} 1 \\ 0 \\ -2 \end{bmatrix}$$

$$v := \begin{bmatrix} -4 \\ 2 \\ -5 \end{bmatrix}$$

Angle between the vectors u and v

$$\theta := acos\left(\frac{u \cdot v}{|u| \cdot |v|}\right)$$

$\theta = 1.159$

$\theta = 66.422 \cdot deg$

For orthogonal vectors

$$x := \begin{bmatrix} 1 \\ 0 \\ 0 \end{bmatrix}$$

$$y := \begin{bmatrix} 0 \\ 1 \\ 0 \end{bmatrix}$$

$x \cdot y = 0$

$|x| \cdot |y| = 1$

$$\Theta := acos\left(\frac{x \cdot y}{|x| \cdot |y|}\right)$$

$\Theta = 1.571$

$\Theta = 90 \cdot deg$

Exhibit 6-49

To convert decimal degrees (or hours) to hours, minutes, and seconds

Example:
    Convert 66.422 degrees to degrees, minutes, and seconds

Step 1
Take the fractional part and multiply by 60

$T := 66.422$

To get just the decimal part of a positive number x, define:

$\text{mantissa}(x) := x - \text{floor}(x)$

$\text{mantissa}(T) := T - \text{floor}(T)$

Or

$M := T - \text{floor}(T)$ $M = 0.422$

$R := M \cdot 60$ $R = 25.32$

The integer part of R is the number of minutes

$\text{min} := \text{floor}(R)$

$\text{min} = 25$

Step 2
Take the fractional part of R and multiply by 60

$S := R - \text{floor}(R)$

$S = 0.32$ $Y := S \cdot 60$

$Y = 19.2$ number of seconds

So the answer is

Degrees $\text{floor}(T) = 66$

Minutes $\text{floor}(R) = 25$

Seconds $\text{floor}(Y) = 19$

Exhibit 6-50

The vector sum operator adds the elements of a vector, as in Exhibit 6-51. Another way to do this is to dot a vector of 1s with the vector containing the values to be summed.

If a vector has three elements and represents a vector in space (where the elements are the x, y, and z components), the magnitude of the vector represents the physical length of the vector. We see here (Exhibit 6-52) how to use the magnitude of a vector to produce a unit vector—a vector that points in the same direction as the original vector and has a length of 1.

The cross-product operator is the second type of multiplication of two vectors. This type of vector multiplication creates a vector as its result (see Exhibits 6-53 to 6-55). We saw that the dot product has a special meaning for 3-vectors, but can be used with vectors with any number of elements as long as both vectors to be multiplied have the same number of elements. The cross-product, on the other hand, is defined only for 3-vectors. It represents "A*B*sin(theta)." Some people associate a mnemonic called the *right-hand rule* with the cross-product. It imagines that if you twist the fingers of the right hand as though to turn from the first vector toward the vector it is being crossed with, the thumb will point in the direction of the resultant vector. In other words, for vectors "A" and "B" in a plane, the resultant of "A × B" will be perpendicular to the plane containing "A" and "B." In physics, several important quantities such as the angular momentum vector are defined in terms of the cross-product (or vector product).

$$a := \begin{bmatrix} 1.45 \\ 2.57 \\ 3.33 \\ 67 \\ 22 \end{bmatrix}$$

Sum the elements of this vector

$\Sigma a = 96.35$

$$i := \begin{bmatrix} 1 \\ 1 \\ 1 \\ 1 \\ 1 \end{bmatrix}$$

Find the sum using the Dot Product

$a \cdot i = 96.35$

Exhibit 6-51

Unit Vector

$$V := \begin{bmatrix} 3 \\ 2 \\ 1.5 \end{bmatrix}$$

$|V| = 3.905$

$u := \dfrac{V}{|V|}$

$$u = \begin{bmatrix} 0.768 \\ 0.512 \\ 0.384 \end{bmatrix}$$

$|u| = 1$ This shows that u is truly a unit vector

Exhibit 6-52

The Cross Product (or vector product)

$$V := \begin{bmatrix} 1 \\ 2 \\ 3 \end{bmatrix}$$

$$V1 := \begin{bmatrix} .5 \\ .5 \\ .5 \end{bmatrix}$$

$$V \times V1 = \begin{bmatrix} -0.5 \\ 1 \\ -0.5 \end{bmatrix}$$

Notice

$$x := \begin{bmatrix} 1 \\ 0 \\ 0 \end{bmatrix}$$

$$y := \begin{bmatrix} 0 \\ 1 \\ 0 \end{bmatrix}$$

Dot Product

$$x \cdot y = 0$$

But,
The Cross Product

$$x \times y = \begin{bmatrix} 0 \\ 0 \\ 1 \end{bmatrix} \quad \text{a vector in the z-direction}$$

Exhibit 6-53

Using the symbolic processor we see how the cross product is calculated.

$$\begin{bmatrix} A0 \\ A1 \\ A2 \end{bmatrix} \times \begin{bmatrix} B0 \\ B1 \\ B2 \end{bmatrix}$$

$$\begin{bmatrix} A1 \cdot B2 - A2 \cdot B1 \\ A2 \cdot B0 - A0 \cdot B2 \\ A0 \cdot B1 - A1 \cdot B0 \end{bmatrix}$$

Exhibit 6-54

Angular momentum is "r × p" (i.e., the cross product of a vector "r" times a vector "p"), where "r" is the vector of distance from a point and "p" is the momentum vector at the end of "r." See Exhibit 6-56 for a calculation of the angular momentum and rotational kinetic energy of the earth.

Some things you can do with arrays

Now that you have become familiar with some of the tools for working with vectors and matrices, we will look at some examples of how these can be used to do real work. You should keep in mind that these are by no means all the tools that Mathcad has available for such work. These are some of the tools that will be most useful in getting started. You will also want to try using some of these techniques for your own work.

Small business.   Assume you own a small business (a restaurant, for example) and you create a matrix for your recordkeeping. Assume also, for this example, that there is no tax and there are no tips. Then your matrix might look something like the one shown in Exhibit 6-57. This matrix has 10 columns and 9 rows. Assume that each row represents one setting at a table and each column represents a particular item on the menu. The first couple of columns might represent the appetizers; the next few columns, the entrées; and the last columns, the desserts and beverages. A one-column vector gives the prices associated with each item on the menu. The multiplication of the food matrix times the price vector gives another vector which lists the total for each check (not including any tax and tip). It is early in the day, so there are only four totals. A realistic matrix for a real restaurant would have many more rows and columns—representing many more table servings and many more menu items.

Markov chain.   Markov chain models are commonly used in many types of research about demographics, marketing, and other fields. A *Markov matrix* is a square matrix in which the elements are transition probabilities (the elements are all nonnegative and the sum of the elements in each row equals 1). Each row of elements represents a set of probabilities that a certain event

$$u := \begin{bmatrix} 1 \\ 0 \\ -2 \end{bmatrix}$$

$$v := \begin{bmatrix} -4 \\ 2 \\ -5 \end{bmatrix}$$

Angle between the vectors u and v using the cross product

$$\theta := \operatorname{asin}\left(\frac{|\, u \times v \,|}{|\, u \,| \cdot |\, v \,|}\right)$$

$$\theta = 1.159$$

$$\theta = 66.422 \cdot \deg$$

For orthogonal vectors

$$x := \begin{bmatrix} 1 \\ 0 \\ 0 \end{bmatrix}$$

$$y := \begin{bmatrix} 0 \\ 1 \\ 0 \end{bmatrix}$$

$$|\, x \times y \,| = 1$$

$$|\, x \,| \cdot |\, y \,| = 1$$

$$x \times y = \begin{bmatrix} 0 \\ 0 \\ 1 \end{bmatrix}$$ So the resultant vector is in the z-direction

$$\Theta := \operatorname{asin}\left(\frac{|\, x \times y \,|}{|\, x \,| \cdot |\, y \,|}\right)$$

$$\Theta = 1.571$$

$$\Theta = 90 \cdot \deg$$

Exhibit 6-55

$$f := \frac{1}{86400 \cdot \sec}$$

$$r := 6.2 \cdot 10^6 \cdot m$$

$$m := 5.98 \cdot 10^{24} \cdot kg$$

$$\omega := 2 \cdot \pi \cdot f$$

$$\omega = 7.272 \bullet 10^{-5} \bullet s^{-1}$$

$$I := \frac{2}{5} \cdot m \cdot r^2$$

$$I = 9.195 \bullet 10^{37} \bullet kg \bullet m^2$$

For Earth

Angular Momentum is r x p (where r is the radius vector and p is the momentum vector) or alternatively I*ω (where I is the moment of inertia and ω is the angular velocity ω=2*π*f)

$$I \cdot \omega = 6.687 \bullet 10^{33} \bullet kg \bullet m^2 \bullet s^{-1} \qquad \text{angular momentum}$$

Rotational Kinetic Energy (K)

$$K := \frac{1}{2} \cdot I \cdot \omega^2$$

$$K = 2.431 \bullet 10^{29} \bullet joule$$

Exhibit 6-56

---

might occur in a specific way (the probabilities of a buyer choosing various specific brands of breakfast cereal, for example). Since the row contains the probabilities for each possible outcome, the total of the probabilities must equal 1. It is assumed that the probabilities stay the same from one stage of a Markov chain process to the next and that the state of the process after each stage (together with the transition probabilities) determines what will happen at the next stage. In Exhibit 6-58 we see a simple example of how the purchases of a consumer product could be described by such a model. The initial matrix contains the transition probabilities (presumably

$$A := \begin{bmatrix} .70 & .20 & .10 \\ .15 & .65 & .20 \\ .09 & .21 & .69 \end{bmatrix}$$

$$A^2 = \begin{bmatrix} 0.53 & 0.29 & 0.18 \\ 0.22 & 0.49 & 0.28 \\ 0.16 & 0.3 & 0.53 \end{bmatrix}$$

$$\begin{bmatrix} 0 & 0 & 1 & 0 & 2 & 0 & 0 & 0 & 0 & 0 \\ 0 & 0 & 0 & 1 & 1 & 0 & 0 & 0 & 0 & 2 \\ 1 & 0 & 0 & 0 & 0 & 0 & 0 & 0 & 0 & 1 \\ 1 & 2 & 0 & 0 & 0 & 3 & 0 & 0 & 1 & 1 \\ 0 & 0 & 0 & 0 & 0 & 0 & 0 & 0 & 0 & 0 \\ 0 & 0 & 0 & 0 & 0 & 0 & 0 & 0 & 0 & 0 \\ 0 & 0 & 0 & 0 & 0 & 0 & 0 & 0 & 0 & 0 \\ 0 & 0 & 0 & 0 & 0 & 0 & 0 & 0 & 0 & 0 \\ 0 & 0 & 0 & 0 & 0 & 0 & 0 & 0 & 0 & 0 \end{bmatrix} \cdot \begin{bmatrix} 2.50 \\ 1.50 \\ 2.20 \\ 6.00 \\ 7.95 \\ 11.00 \\ 7.95 \\ 3.50 \\ 2.00 \\ 1.00 \end{bmatrix} = \begin{bmatrix} 18.1 \\ 15.95 \\ 3.5 \\ 41.5 \\ 0 \\ 0 \\ 0 \\ 0 \\ 0 \end{bmatrix}$$

$$A^3 = \begin{bmatrix} 0.4301 & 0.3325 & 0.2346 \\ 0.254 & 0.425 & 0.3162 \\ 0.202 & 0.3366 & 0.4392 \end{bmatrix}$$

$$A^{16} = \begin{bmatrix} 0.2782 & 0.356 & 0.3227 \\ 0.2767 & 0.3544 & 0.3215 \\ 0.271 & 0.3472 & 0.315 \end{bmatrix}$$

$$A^{30} = \begin{bmatrix} 0.2652 & 0.3396 & 0.308 \\ 0.264 & 0.338 & 0.3066 \\ 0.2586 & 0.3312 & 0.3004 \end{bmatrix}$$

Exhibit 6-57   **Exhibit 6-58**

estimated from data on actual purchases) for the popular brands of this product. The top row consists of the probabilities for purchasing brands A, B, and C by the people who were observed to initially purchase brand A. The middle row gives the probabilities for those who initially purchased brand B, and the bottom row gives the probabilities for brand C purchasers. Thus 70 percent of those who initially purchased brand A are expected to purchase it again on the next round of purchases, but 20 percent of them will purchase brand B and 10 percent will purchase brand C. Raising the matrix to successively higher powers will project the proportions of people from the initial classification who buy each product at successive stages, assuming that the transition probabilities stay the same. Notice that after many stages, the proportions tend to approach an equilibrium where they change very little in successive stages.

Pauli matrices. *Pauli matrices* are simple matrices, described by the physicist Wolfgang Pauli, that have some interesting properties that make them very important in quantum mechanics. See Exhibit 6-59.

Fourier analysis and fast Fourier transform. If your work involves Fourier analysis, you will be happy to know that Mathcad has several powerful procedures for fast Fourier transform. The Professional edition of Mathcad is especially rich in these techniques. Exhibits 6-60 and 6-61

$$\Sigma1 := \begin{bmatrix} 0 & 1 \\ 1 & 0 \end{bmatrix} \qquad \Sigma2 := \begin{bmatrix} 0 & -i \\ i & 0 \end{bmatrix} \qquad \Sigma3 := \begin{bmatrix} 1 & 0 \\ 0 & -1 \end{bmatrix}$$

Some interesting properties of Pauli Matrices include

$$\Sigma1^2 = \begin{bmatrix} 1 & 0 \\ 0 & 1 \end{bmatrix} \qquad \Sigma2^2 = \begin{bmatrix} 1 & 0 \\ 0 & 1 \end{bmatrix} \qquad \Sigma3^2 = \begin{bmatrix} 1 & 0 \\ 0 & 1 \end{bmatrix}$$

And

$$\Sigma1 \cdot \Sigma2 = \begin{bmatrix} 1i & 0 \\ 0 & -1i \end{bmatrix} \qquad \text{Which is equivalent to} \quad i \text{ times } \Sigma3 \qquad i \cdot \Sigma3 = \begin{bmatrix} 1i & 0 \\ 0 & -1i \end{bmatrix}$$

$$\Sigma2 \cdot \Sigma3 = \begin{bmatrix} 0 & 1i \\ 1i & 0 \end{bmatrix} \qquad\qquad i \text{ times } \Sigma1 \qquad i \cdot \Sigma1 = \begin{bmatrix} 0 & 1i \\ 1i & 0 \end{bmatrix}$$

$$\Sigma3 \cdot \Sigma1 = \begin{bmatrix} 0 & 1 \\ -1 & 0 \end{bmatrix} \qquad\qquad i \text{ times } \Sigma2 \qquad i \cdot \Sigma2 = \begin{bmatrix} 0 & 1 \\ -1 & 0 \end{bmatrix}$$

$$\Sigma2 \cdot \Sigma1 = \begin{bmatrix} -1i & 0 \\ 0 & 1i \end{bmatrix} \qquad\qquad -i \text{ times } \Sigma3 \qquad -i \cdot \Sigma3 = \begin{bmatrix} -1i & 0 \\ 0 & 1i \end{bmatrix}$$

$$\Sigma3 \cdot \Sigma2 = \begin{bmatrix} 0 & -1i \\ -1i & 0 \end{bmatrix} \qquad\qquad -i \text{ times } \Sigma1 \qquad -i \cdot \Sigma1 = \begin{bmatrix} 0 & -1i \\ -1i & 0 \end{bmatrix}$$

$$\Sigma1 \cdot \Sigma3 = \begin{bmatrix} 0 & -1 \\ 1 & 0 \end{bmatrix} \qquad\qquad -i \text{ times } \Sigma2 \qquad -i \cdot \Sigma2 = \begin{bmatrix} 0 & -1 \\ 1 & 0 \end{bmatrix}$$

$$\Sigma1 \cdot \Sigma2 + \Sigma2 \cdot \Sigma1 = \begin{bmatrix} 0 & 0 \\ 0 & 0 \end{bmatrix}$$

$$\Sigma2 \cdot \Sigma3 + \Sigma3 \cdot \Sigma2 = \begin{bmatrix} 0 & 0 \\ 0 & 0 \end{bmatrix}$$

$$\Sigma3 \cdot \Sigma1 + \Sigma1 \cdot \Sigma3 = \begin{bmatrix} 0 & 0 \\ 0 & 0 \end{bmatrix}$$

Exhibit 6-59 Pauli matrices.

$$v := \begin{bmatrix} 0 \\ 1 \\ 2 \\ 3 \\ 4 \\ 0 \\ 1 \\ 2 \\ 3 \\ 4 \\ 0 \\ 1 \\ 2 \\ 3 \\ 4 \\ 0 \end{bmatrix}$$

$$v := \begin{bmatrix} 1 \\ 0 \\ 0 \\ 3 \\ 4 \\ 5 \\ 0 \\ 0 \\ 4 \\ 5 \\ 6 \\ 0 \\ 0 \\ 5 \\ 6 \\ 7 \end{bmatrix}$$

$$\text{FFT}(v) = \begin{bmatrix} 1.875 \\ -0.136 + 8.544 \cdot 10^{-4}\,i \\ -0.188 + 0.011i \\ -0.817 + 0.203i \\ 0.125 - 0.125i \\ -1.699 \cdot 10^{-4} - 0.135i \\ -0.187 - 0.364i \\ 0.203 + 0.162i \\ 0.125 \end{bmatrix}$$

$$\text{FFT}(v) = \begin{bmatrix} 2.875 \\ -2.619 \cdot 10^{-4} + 0.394i \\ 0.018 + 0.398i \\ -0.313 + 1.545i \\ -0.188 - 0.313i \\ -0.062 - 0.016i \\ 0.107 + 0.398i \\ -0.375 - 0.167i \\ -0.25 \end{bmatrix}$$

Exhibit 6-60   **Exhibit 6-61**

give a couple of simple examples of how to use one of these transforms, given a vector of numerical values. A full description of how to use all the available transforms is beyond the scope of this book. The Professional edition of Mathcad also includes the capability to treat problems involving wavelets.

Exhibit 6-62 consists of a real-life example developed by one of the authors, based on a

This example computes Fourier coefficients for a composite piston motion function and generates an approximating Fourier polynomial. The piston motion is the first 90 degrees of rotation rotated or mirrored and repeated for the full 360 degrees. The first 90 degrees of motion is truncated at b degrees and a straight line is used for the transition.

$c := 120$ Connecting rod length

$s := 68$ Stroke

$L := 180$ positive endpoint of periodic interval:

$b := 70$ The length of repeating portion of the curve starting from -180 degrees

$r := \dfrac{s}{2}$ One half of the stroke

$$l2(x) := r + c + r \cdot \cos(x \cdot \deg) - \sqrt{c^2 - (r \cdot \sin(x \cdot \deg))^2}$$ Crank and Rod Motion

$$dr(x) := r - c - r \cdot \cos((x - 180) \cdot \deg) + \sqrt{c^2 - (r \cdot \sin((x - 180) \cdot \deg))^2}$$ New Piston motion

$a1 := -L + b$ End of the repeating portion of the curve $a1 = -110$

$d1 := l2(a1)$ Lift at a1 and a4 $d1 = 26.703$

$a2 := -b$ Start of second portion of the curve $a2 = -70$

$d2 := dr(a2)$ Lift at a2 and a3 $d2 = 41.297$

$m1 := \dfrac{d2 - d1}{a2 - a1}$ Slope of a line from a1 to a2 $m1 = 0.365$

$b1 := d1 - m1 \cdot a1$ Intercept of a line from a1 to a2 $b1 = 66.838$

$a3 := -a2$ End of third portion of the curve $a3 = 70$

$a4 := -a1$ Start of fourth portion of the curve $a4 = 110$

$m2 := \dfrac{d2 - d1}{a3 - a4}$ Slope of a line from a3 to a4 $m2 = -0.365$

$b2 := d1 - m2 \cdot a4$ Intercept of a line from a3 to a4 $b2 = 66.838$

Enter a function that is periodic on interval $-L \le x \le L$:

$f(x) := \begin{vmatrix} l2(x) & \text{if } -L \le x \le a1 \\ m1 \cdot x + b1 & \text{if } a1 \le x \le a2 \\ dr(x) & \text{if } a2 \le x \le a3 \\ m2 \cdot x + b2 & \text{if } a3 \le x \le a4 \\ l2(x) & \text{if } a4 \le x \le L \end{vmatrix}$

Order of Fourier series approximation: $N := 5$

Exhibit 6-62 Computing Fourier coefficients.

Program to compute Fourier coefficients:

$$FC(f, N, L) := \begin{vmatrix} R^{\langle 0 \rangle} \leftarrow \begin{bmatrix} \dfrac{1}{2 \cdot L} \cdot \displaystyle\int_{-L}^{L} f(x)\, dx \\ 0 \end{bmatrix} \\ \text{for } n \in 1 .. N \\ \qquad R^{\langle n \rangle} \leftarrow \begin{bmatrix} \dfrac{1}{L} \cdot \displaystyle\int_{-L}^{L} f(x) \cdot \cos\left(\dfrac{n \cdot \pi \cdot x}{L}\right) dx \\ \dfrac{1}{L} \cdot \displaystyle\int_{-L}^{L} f(x) \cdot \sin\left(\dfrac{n \cdot \pi \cdot x}{L}\right) dx \end{bmatrix} \\ (R)^T \end{vmatrix}$$

Computed Fourier coefficients:

$$res := FC(f, N, L) \qquad A := res^{\langle 0 \rangle} \qquad B := res^{\langle 1 \rangle}$$

Nth Fourier polynomial:

$$p(x) := A_0 + \left[\sum_{n=1}^{N} \left(A_n \cdot \cos\left(\frac{n \cdot \pi \cdot x}{L}\right) + B_n \cdot \sin\left(\frac{n \cdot \pi \cdot x}{L}\right) \right) \right]$$

$n := 36$ Number of steps across the plot from -180 to 180

$x := -L, -L + \dfrac{L}{n} .. L$ range variable from -180 degrees to 180 degrees in n steps

Unadjusted polynomial curve results...

$p(0) = 67.777$ Maximum Lift

$p(180) = 0.223$ Minimum Lift $p(-180) = 0.223$

The lift curve must be factored so that the minimum lift is exactly zero,
and the maximum lift is exactly equal to the stroke $s = 68$.

$$ll(x) := (p(x) - p(180)) \cdot \frac{s}{p(0) - p(180)} \qquad \text{New lift function using the polynomial fit}$$

Maximum Lift $ll(0) = 68$

Minimum Lift $ll(180) = 0$

The equation proves to produce the correct lift.

Exhibit 6-62 (*Continued*)

Lift

$$\text{Lift} \quad \frac{l1(x)}{l2(x)}$$

Degrees of Crank Rotation

— Fourier polynomial
— Crank and Rod

Lift Curve

Velocity

$$v1(x) := \frac{d}{dx} l1(x) \qquad v2(x) := \frac{d}{dx} l2(x)$$

$$\text{Velocity} \quad \frac{v1(x)}{v2(x)}$$

Degrees of Crank Rotation

— Fourier Polynomial
— Crank and Rod

Velocity Curve

Acceleration

$$a1(x) := \frac{d^2}{dx^2} l1(x) \qquad a2(x) := \frac{d^2}{dx^2} l2(x)$$

$$\text{Acceleration} \quad \frac{a1(x)}{a2(x)}$$

Degrees of Crank Rotation

— Fourier Polynomial
— Crank and Rod

Acceleration Curve

Jerk

$$j1(x) := \frac{d^3}{dx^3} l1(x) \qquad j2(x) := \frac{d^3}{dx^3} l2(x)$$

$$\text{Jerk} \quad \frac{j1(x)}{j2(x)}$$

Degrees of Rotation

— Fourier Polynomial
— Crank and Rod

Jerk Curve

Exhibit 6-62 (*Continued*)

$$\text{Incomes} := \begin{bmatrix} 10000 \\ 55000 \\ 32000 \\ 11000 \\ 10000000 \\ 76000 \\ 33500 \\ 25000 \\ 120000 \\ 57000 \\ 5000 \\ 13000 \\ 44000 \\ 22000 \\ 90000 \end{bmatrix}$$

$\text{mean}(\text{Incomes}) = 706233$

$\text{median}(\text{Incomes}) = 33500$

Notice how the one very large income skews the mean
and standard deviation for our income sample

$\text{stdev}(\text{Incomes}) = 2484065$

Exhibit 6-63

Mathcad "QuickSheet," for computing Fourier coefficients for a polynomial that will fit between two other curves as part of the design for a component of an engine. The actual example shown here is a first approximation to a solution. The correct solution will involve some additional fine tuning.

Basic statistics

Mathcad has a wealth of powerful statistical techniques built in, and you can easily define additional functions for your own work. Here we introduce three of the most widely used statistical functions:

- The mean
- The median
- The standard deviation

Given a vector of values, Mathcad makes it easy to calculate these basic statistics as seen in Exhibit 6-63.

Linear regression.   A commonly used technique for analyzing your results is the straight-line regression. We give an example (Exhibit 6-64) of how, given vectors of x values and y values,

---

$i := 0.. 19$

$x := (1 \quad 2 \quad 3 \quad 4 \quad 5 \quad 6 \quad 7 \quad 8 \quad 9 \quad 10 \quad 11 \quad 12 \quad 13 \quad 14 \quad 15 \quad 16 \quad 17 \quad 18 \quad 19 \quad 20)^T$

$y := (2.3 \quad 4.2 \quad 3 \quad 5.5 \quad 6.6 \quad 7.0 \quad 4.5 \quad 9.0 \quad 35 \quad 15.1 \quad 17.0 \quad 18 \quad 16.4 \quad 11.3 \quad 26 \quad 15 \quad 22 \quad 19.7 \quad 30 \quad 16.9)^T$

Compute

$m := \text{slope}(x, y)$

$b := \text{intercept}(x, y)$

Then the least-squares line is given by

$f(d) := m \cdot d + b$

the graph shows the data and the least squares regression line.

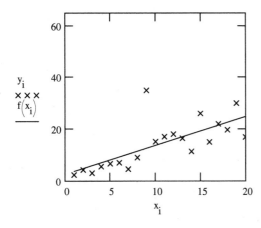

Exhibit 6-64

Translation and Rotation of 2-D Axes

Situation:
The axes are translated to a new origin (h,k) and then rotated through an angle θ, the equations
of the transformation are then:

x = x' * cos θ - y' * sin θ + h

y = x' * sin θ + y' * cos θ + k

$i := 0.. 6$

$h := 6$

$k := 2.35$

$θ := 1.2$

$$X := \begin{bmatrix} -1 \\ -.5 \\ .23 \\ .34 \\ 3 \\ 10 \\ 15 \end{bmatrix} \qquad Y := \begin{bmatrix} -2 \\ .2 \\ 3 \\ 4 \\ 7 \\ 11 \\ 15 \end{bmatrix}$$

$x_i := X_i \cdot \cos(θ) - Y_i \cdot \sin(θ) + h$

$y_i := X_i \cdot \sin(θ) + Y_i \cdot \cos(θ) + k$

$$x = \begin{bmatrix} 7.502 \\ 5.632 \\ 3.287 \\ 2.395 \\ 0.563 \\ -0.629 \\ -2.545 \end{bmatrix} \qquad y = \begin{bmatrix} 0.693 \\ 1.956 \\ 3.651 \\ 4.116 \\ 7.683 \\ 15.656 \\ 21.766 \end{bmatrix}$$

Exhibit 6-65 Translation and rotation of 2D axes.

to do such a linear regression in Mathcad. Other types of regressions (nonlinear, multivariable, etc.) are also easy to do using Mathcad's built-in statistical capabilities, but we cannot cover all of these here. The Mathcad Treasury has a wealth of information on using these.

Scatter diagrams. It is often desirable to produce a plot that shows the points produced by vectors of x values and y values. This scatter diagram is demonstrated in Exhibit 6-64, along with the linear regression. These points might represent stock prices over time, heights and weights of students, or other data. Such a scatter diagram might be drawn to get an "eyeball" view of the relationships between two variables before doing a regression.

Translation and rotation of coordinates. Consider a situation where you are working with a specific x-y coordinate system. Assume that a new coordinate system with X and Y axes is then created that has been moved (translated) to a new location ($x = h$, $y = k$) and rotated through an angle θ. An example of such a situation is given in Exhibit 6-65. The problem is to take a vector "X" of x values in the new coordinate system, and a vector "Y" of y values in the new coordinate system, and calculate the coordinates with respect to the original coordinate system. Once again, you see here how an important, but tedious, calculation can be reduced to a simple Mathcad procedure.

RMS deviation. The root-mean-square (RMS) deviation measure is used in statistics and the physical sciences. Mathcad allows you to readily calculate it (Exhibit 6-66). In Exhibit 6-67, we have filled a vector with random numbers and used it to compute another example of RMS deviation. Mathcad provides several ways to produce random numbers. The random numbers generated in this example represent a uniform distribution. The arguments tell Mathcad to fill

---

$$X := \begin{bmatrix} -3 \\ 5 \\ -7 \\ 9 \\ 11 \\ 4 \\ -6 \end{bmatrix}$$

$$RMS(x) := \sqrt{\frac{1}{length(x)} \cdot (x \cdot x)}$$

$$RMS(X) = 6.939$$

Exhibit 6-66

$$A := \text{runif}(100, 0, 1)$$

$$A = $$

	0
0	$1.268 \cdot 10^{-3}$
1	0.193
2	0.585
3	0.35
4	0.823
5	0.174
6	0.71
7	0.304
8	0.091
9	0.147
10	0.989
11	0.119
12	$8.923 \cdot 10^{-3}$
13	0.532
14	0.602
15	0.166

$$RMS(x) := \sqrt{\frac{1}{\text{length}(x)} \cdot (x \cdot x)}$$

$$RMS(A) = 0.58$$

Exhibit 6-67

vector "A" with 100 random numbers with values between 0 and 1. Other Mathcad generators of random numbers have distributions suitable for many types of problems [normally distributed, having the F distribution, having the chi-squared (χ^2) distribution, etc.].

Lottery payout.   A lottery promises payments to the winner of a specific amount for a specific number of years. The present value of such a series of lottery payouts is determined in Exhibit 6-68. Each annual payment is discounted for the time when it is payable, in order to compute a value that represents what such a series of payments is currently worth—adjusted for when each payment is scheduled to be paid.

Bond yield.   An investor is interested in purchasing a bond that would provide a series of annual interest payments and then return the invested principal on the bond's maturity date.

A lottery winner is scheduled to get $2 million immediately, and $2 million at the end of each of the next 19 years.

Assume that the relevant interest rate is 7%.

What is the present value of this contract?

The main thing to remember is that each payment needs to be discounted to reflect when it is to be received.

The discount factors are

$$\frac{1}{(1+i)^n}$$ where n is the year the payment is received.

The initial payment is not discounted, because it is to be paid out immediately.

$i := .07$

$$PV := 2 + \sum_{n=1}^{19} \frac{2}{(1+i)^n}$$

So the answer (in millions of dollars) is that the Present Value

$PV = 22.671$

So the state lottery commission could expect to pay some company (such as an insurance company or pension administrator) approximately $22.671 million to make the payments to the winner—assuming that the state does not want to do the administrative work itself. The exact amount would depend on how much the administrator needed to add for its administrative expenses and a reasonable profit margin. Presumably, the contract would be bid on by some potential administrators and so competitive bidding would determine the precise amount.

Exhibit 6-68

$R := .06$

A bond pays annual interest of $1350.
It will mature in 12 years for $20,000.
What is the bond worth (net present value or purchase
price) if you want to earn a 7.5% yield to maturity?

$$cf := \begin{bmatrix} -1200 \\ 100 \\ 200 \\ 300 \\ 400 \\ 500 \\ 600 \end{bmatrix}$$

$$PVI := \sum_{n=1}^{12} \frac{1350}{(1+.075)^n}$$ Present value of interest payments

$$PVMV := \frac{20000}{(1+.075)^{12}}$$ Present value of maturity value

$$PV := \sum_{i=0}^{6} \frac{cf_i}{(1+R)^i}$$

$PV := PVI + PVMV$ Total present
value

$PV = 18839.71$ $PV = 437.668$

Exhibit 6-69   **Exhibit 6-70**

The investor has the goal of achieving a certain target rate of return on the investment. This example shows how you can determine the amount such an investor would be willing to pay for such a bond. See Exhibit 6-69.

Value of a series of cash flows.   This exhibit shows how one would evaluate the net present value of a series of cash flows such as might represent the investment and the return on investment for a new business, or some new piece of equipment that will improve the profitability of an existing business. For example, a purchase of a new computer and software might initially involve an outflow of money, but the investment is expected to result in new revenues or lower expenses. The outflows (investments) are negative numbers, and the returns from the investment are positive. The calculation of net present values is based on an interest rate—your current cost of capital. Exhibits 6-70 and 6-71 are examples showing different interest rates.

Vectorizing

For some problems it is desirable to use a Mathcad capability that works with vectors and matrices but creates different results than those from the normal matrix, vector, and range variables. Vectorizing can be used to multiply vectors element by element, or to operate on

$i := 0 .. 6$

$r := .07$

$$A := \begin{bmatrix} 1 & 4 & 3 \\ 2 & 5 & 4 \\ 1 & -3 & -2 \end{bmatrix}$$

$$a := \begin{bmatrix} -15000 \\ 200 \\ 600 \\ 1200 \\ 4500 \\ 7000 \\ 13000 \end{bmatrix}$$

Matrix multiplication

$$A \cdot A = \begin{bmatrix} 12 & 15 & 13 \\ 16 & 21 & 18 \\ -7 & -5 & -5 \end{bmatrix}$$

Vectorized multiplication (each element of the first matrix multiplied by the corresponding element of the second matrix)

$A = 1 \cdot A$

$$A := \sum_i \frac{a_i}{(1+r)^i}$$

$$\overrightarrow{A \cdot A} = \begin{bmatrix} 1 & 16 & 9 \\ 4 & 25 & 16 \\ 1 & 9 & 4 \end{bmatrix}$$

Exhibit 6-71   **Exhibit 6-72**

---

vectors without using range variables. Some examples are shown below:

- Exhibit 6-72 demonstrates the difference from matrix multiplication.
- Exhibits 6-73 through 6-76 show examples of calculations using range variables and vectorizing.
- Exhibit 6-77 shows how vectorizing can be used for dividing vectors or computing compound annual growth rates.

Redoing Lorentz contraction. In an earlier section of this book, you saw how objects moving at very high speeds (as compared to the speed of light) undergo Lorentz contraction. Here we show how to do this calculation by vectorizing (see Exhibit 6-78).

Eigenvalues and eigenvectors

If you use eigenvalues and eigenvectors in your work, you will appreciate Mathcad's powerful capabilities to compute them, especially in the Professional version. Given a matrix "A" as argument, "eigenvals" computes the eigenvalues. You can then use "eigenvec" to compute the

Range variable (instead of vectorize)

$a := 0 .. 7$

$$V := \begin{bmatrix} 0 \\ 1 \\ \dfrac{\pi}{2} \\ 2.2 \\ 3.22 \\ 4.2 \\ 5 \\ 7 \end{bmatrix}$$

$$V := \begin{bmatrix} 0 \\ 1 \\ \dfrac{\pi}{2} \\ 2.2 \\ 3.22 \\ 4.2 \\ 5 \\ 7 \end{bmatrix}$$

$E_a := e^a$

Vectorize

$$E = \begin{bmatrix} 1 \\ 2.718 \\ 7.389 \\ 20.086 \\ 54.598 \\ 148.413 \\ 403.429 \\ 1.097 \cdot 10^3 \end{bmatrix}$$

$$\overrightarrow{e^V} = \begin{bmatrix} 1 \\ 2.718 \\ 4.81 \\ 9.025 \\ 25.028 \\ 66.686 \\ 148.413 \\ 1.097 \cdot 10^3 \end{bmatrix}$$

Exhibit 6-73   **Exhibit 6-74**

---

corresponding eigenvectors for the various eigenvalues, or Mathcad Professional users can use "eigenvecs" to compute all eigenvectors at once (Exhibits 6-79 and 6-80).

Sorting vectors

A common need in working with arrays of numbers is a way to sort the numbers. Mathcad provides "sort" and "reverse," which take a vector name as argument. Exhibit 6-81 shows how "sort" sorts the elements of the vector into ascending order and "reverse" reverses the order

Range Variable

$$a := 0 \,.. \, 7$$

$$V_a := 1.05^a$$

$$V = \begin{bmatrix} 1 \\ 1.05 \\ 1.103 \\ 1.158 \\ 1.216 \\ 1.276 \\ 1.34 \\ 1.407 \end{bmatrix}$$

$$A := \begin{bmatrix} 1.2 \\ 12 \\ 11 \\ 15 \\ 17 \end{bmatrix}$$

$$B := \begin{bmatrix} 6 \\ 22 \\ 15 \\ 31 \\ 21 \end{bmatrix}$$

Vectorize

$$V := \begin{bmatrix} 0 \\ 1 \\ 2 \\ 3 \\ 4 \\ 5 \\ 6 \\ 7 \end{bmatrix}$$

$$V := \begin{bmatrix} 0 \\ 1 \\ \dfrac{\pi}{2} \\ 2.2 \\ 3.22 \\ 4.2 \\ 5 \\ 7 \end{bmatrix}$$

Vectorized Division

$$\dfrac{\overrightarrow{B}}{A} = \begin{bmatrix} 5 \\ 1.833 \\ 1.364 \\ 2.067 \\ 1.235 \end{bmatrix}$$

Vectorize

Vectorized Compound Annual Growth Rate

$$\overrightarrow{\cos(V)} = \begin{bmatrix} 1 \\ 0.54 \\ 0 \\ -0.589 \\ -0.997 \\ -0.49 \\ 0.284 \\ 0.754 \end{bmatrix}$$

$$\overrightarrow{1.05^V} = \begin{bmatrix} 1 \\ 1.05 \\ 1.103 \\ 1.158 \\ 1.216 \\ 1.276 \\ 1.34 \\ 1.407 \end{bmatrix}$$

$$\overrightarrow{\left(\dfrac{B}{A}\right)^{\frac{1}{5}}} = \begin{bmatrix} 1.38 \\ 1.129 \\ 1.064 \\ 1.156 \\ 1.043 \end{bmatrix}$$

Exhibit 6-75   **Exhibit 6-76**   **Exhibit 6-77**

$c := 300000$

$$V := \begin{bmatrix} 1 \\ 100 \\ 1000 \\ 10000 \\ 100000 \\ 200000 \\ .9 \cdot c \end{bmatrix}$$

$$A := \begin{bmatrix} 1 & 5 \\ 2 & 4 \end{bmatrix}$$

$$A := \sqrt{1 - \frac{\overrightarrow{V^2}}{c^2}}$$

$$\text{eigenvals}(A) = \begin{bmatrix} -1 \\ 6 \end{bmatrix}$$

$$\text{eigenvec}(A, -1) = \begin{bmatrix} -0.928 \\ 0.371 \end{bmatrix}$$

$$\text{eigenvec}(A, 6) = \begin{bmatrix} 0.707 \\ 0.707 \end{bmatrix}$$

$$A = \begin{bmatrix} 1 \\ 1 \\ 1 \\ 0.999 \\ 0.943 \\ 0.745 \\ 0.436 \end{bmatrix}$$

Or

$$\text{eigenvecs}(A) = \begin{bmatrix} 0.928 & 0.707 \\ -0.371 & 0.707 \end{bmatrix}$$

Exhibit 6-78 Vectorized
Lorentz contraction.

Exhibit 6-79

---

of the vector elements. At the bottom of the page, you see that reverse and sort can be used together (nested). The equivalent sorting commands for matrices are "csort" and "rsort," which require an additional argument to tell Mathcad which column ("csort") or row ("rsort") is to be sorted into ascending order. Remember, as usual, that the top row and the leftmost column are designated as row 0 and column 0, respectively.

Reconciliation of a checking account

We have seen some powerful uses of arrays in Mathcad for science, engineering, and finance. Mathcad's array capabilities can also be used for many relatively routine transactions such as household accounting. Exhibit 6-82 shows how you could balance your checkbook using Mathcad.

Exhibits 6-83 and 6-84 give simple examples of fast Fourier transform calculations.

Solving for the currents in a circuit

In Exhibit 6-85 we show a circuit diagram based on one given by A. W. Joshi in his book *Matrices and Tensors in Physics* (Wiley Eastern, 1985). Using Kirchhoff's Laws for electrical circuits, we produced the matrices shown in the exhibit and solved, using matrix methods, for the currents in the circuits. After solving the problem with the numerical values in Exhibit 6-86, we redid it in symbolic form at the bottom of the exhibit before substituting in the numerical values to confirm that the result is the same.

$$A := \begin{bmatrix} 1 & -1 & 1 \\ 1 & 1 & -1 \\ 2 & -1 & 0 \end{bmatrix}$$

$$\text{eigenvals}(A) = \begin{bmatrix} -1 \\ 1 \\ 2 \end{bmatrix}$$

$$\text{eigenvecs}(A) = \begin{bmatrix} -0.169 & 0.577 & 0.707 \\ 0.507 & 0.577 & 0 \\ 0.845 & 0.577 & 0.707 \end{bmatrix}$$

Exhibit 6-80

$$A := \begin{bmatrix} 32.67 \\ 100.00 \\ 24.56 \\ 100.99 \\ 23.56 \\ 55.00 \\ 75.00 \\ 55.00 \\ 23.98 \\ 23.55 \end{bmatrix}$$

sort(A) =

	0
0	23.55
1	23.56
2	23.98
3	24.56
4	32.67
5	55
6	55
7	75
8	100
9	100.99

reverse(A) =

	0
0	23.55
1	23.98
2	55
3	75
4	55
5	23.56
6	100.99
7	24.56
8	100
9	32.67

$$B := \begin{bmatrix} 1 & 2 & 3.5 & 1000 \\ 11 & 4 & 100 & 789 \\ 130 & 8 & 23 & 130 \\ 150.5 & 97 & .78 & 25 \end{bmatrix}$$

$$\text{csort}(B, 3) = \begin{bmatrix} 150.5 & 97 & 0.78 & 25 \\ 130 & 8 & 23 & 130 \\ 11 & 4 & 100 & 789 \\ 1 & 2 & 3.5 & 1 \cdot 10^3 \end{bmatrix}$$

reverse(sort(A)) =

	0
0	100.99
1	100
2	75
3	55
4	55
5	32.67
6	24.56
7	23.98
8	23.56
9	23.55

$$\text{rsort}(B, 3) = \begin{bmatrix} 3.5 & 1 \cdot 10^3 & 2 & 1 \\ 100 & 789 & 4 & 11 \\ 23 & 130 & 8 & 130 \\ 0.78 & 25 & 97 & 150.5 \end{bmatrix}$$

Exhibit 6-81

Balance last statement = LB

Vector of bank fees = FEE

Interest added to the account since last statement = INT

Vector of deposits not shown on last statement = DEP

Vector of checks not cleared through the last statement = CKS

Vector of ATM debits not shown on last statement = ATMDEB

Vector of ATM credits (transfers from savings, etc.) not
   shown on last statement = ATMCRED

Current Balance = CB

$$LB := 1232.57 \qquad FEE := \begin{bmatrix} 2.50 \\ 3.00 \\ 11.00 \end{bmatrix} \qquad INT := 0 \qquad DEP := \begin{bmatrix} 120.00 \\ 50.00 \\ 20.00 \\ 200.00 \\ 55.00 \\ 120.45 \end{bmatrix}$$

$$CKS := \begin{bmatrix} 32.67 \\ 100.00 \\ 24.56 \\ 100.99 \\ 23.56 \\ 55.00 \\ 75.00 \\ 55.00 \\ 23.98 \\ 23.55 \end{bmatrix} \qquad ATMDEB := \begin{bmatrix} 60.00 \\ 100.00 \\ 60.00 \\ 70.00 \\ 40.00 \end{bmatrix} \qquad ATMCRED := \begin{bmatrix} 200.00 \\ 100.00 \\ 400.00 \end{bmatrix}$$

$$CB := (LB - \Sigma FEE) + INT + \Sigma DEP - \Sigma CKS - \Sigma ATMDEB + \Sigma ATMCRED$$

$$CB = 1.637 \cdot 10^3$$

Exhibit 6-82

$$v := \begin{bmatrix} 0 \\ 1 \\ 2 \\ 3 \\ 4 \\ 0 \\ 1 \\ 2 \\ 3 \\ 4 \\ 0 \\ 1 \\ 2 \\ 3 \\ 4 \\ 0 \end{bmatrix}$$

$$FFT(v) = \begin{bmatrix} 1.875 \\ -0.136 + 8.544 \cdot 10^{-4} i \\ -0.188 + 0.011i \\ -0.817 + 0.203i \\ 0.125 - 0.125i \\ -1.699 \cdot 10^{-4} - 0.135i \\ -0.187 - 0.364i \\ 0.203 + 0.162i \\ 0.125 \end{bmatrix}$$

Exhibit 6-83

$$v := \begin{bmatrix} 1 \\ 0 \\ 0 \\ 3 \\ 4 \\ 5 \\ 0 \\ 0 \\ 4 \\ 5 \\ 6 \\ 0 \\ 0 \\ 5 \\ 6 \\ 7 \end{bmatrix}$$

$$\text{FFT}(v) = \begin{bmatrix} 2.875 \\ -2.619 \cdot 10^{-4} + 0.394i \\ 0.018 + 0.398i \\ -0.313 + 1.545i \\ -0.188 - 0.313i \\ -0.062 - 0.016i \\ 0.107 + 0.398i \\ -0.375 - 0.167i \\ -0.25 \end{bmatrix}$$

Exhibit 6-84

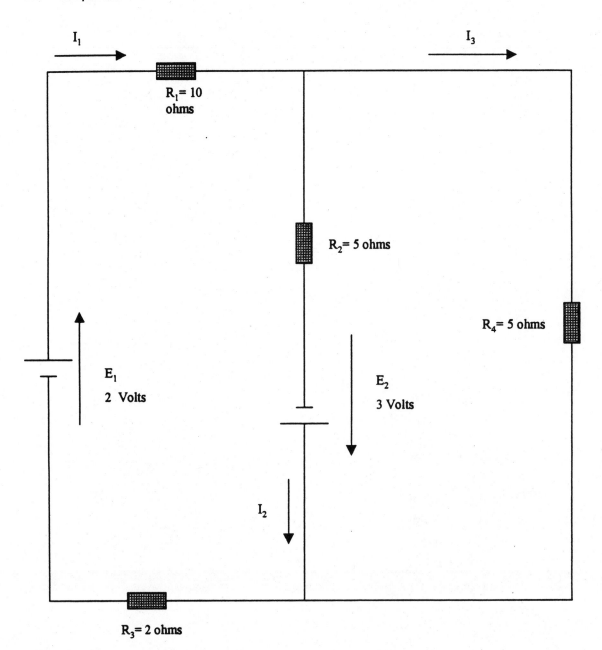

Exhibit 6-85

R1 := 10

R2 := 5

R3 := 2

R4 := 5

E1 := 2

E2 := 3

$$A := \begin{bmatrix} 1 & -1 & -1 \\ 0 & -R2 & R4 \\ R1 + R3 & 0 & R4 \end{bmatrix}$$

$$B := \begin{bmatrix} 0 \\ -E2 \\ E1 \end{bmatrix}$$

$$A^{-1} \cdot B = \begin{bmatrix} 0.241 \\ 0.421 \\ -0.179 \end{bmatrix}$$

$$\begin{bmatrix} 1 & -1 & -1 \\ 0 & -R2 & R4 \\ R1 + R3 & 0 & R4 \end{bmatrix}^{-1} \cdot \begin{bmatrix} 0 \\ -E2 \\ E1 \end{bmatrix}$$

$$\frac{1}{(-R2 \cdot R4 - R1 \cdot R4 - R1 \cdot R2 - R3 \cdot R4 - R3 \cdot R2)} \cdot \begin{bmatrix} -R4 \cdot E2 - E1 \cdot R4 - E1 \cdot R2 \\ -R4 \cdot E2 - E2 \cdot R1 - E2 \cdot R3 - E1 \cdot R4 \\ E2 \cdot R1 + E2 \cdot R3 - E1 \cdot R2 \end{bmatrix}$$

$$\frac{1}{(-R2 \cdot R4 - R1 \cdot R4 - R1 \cdot R2 - R3 \cdot R4 - R3 \cdot R2)} \cdot \begin{bmatrix} -R4 \cdot E2 - E1 \cdot R4 - E1 \cdot R2 \\ -R4 \cdot E2 - E2 \cdot R1 - E2 \cdot R3 - E1 \cdot R4 \\ E2 \cdot R1 + E2 \cdot R3 - E1 \cdot R2 \end{bmatrix} = \begin{bmatrix} 0.241 \\ 0.421 \\ -0.179 \end{bmatrix}$$

Exhibit 6-86

Selected Examples from Various Disciplines

Now that you have become familiar with some of Mathcad's most useful capabilities, this chapter explores some additional examples of problems that illustrate how to put Mathcad to work in a number of disciplines.

Compound Annual Growth Rates Calculation

For evaluating investment returns and growth rates of businesses or companies, it is often desirable to compute the compound annual growth rate. In Exhibit 7-1 we calculate the compound annual growth rate over 5 years for a mutual fund. For this example, the time span is 5 years. If a different number of years were used (e.g., 6 years) the $\frac{1}{5}$ power would change (to $\frac{1}{6}$ for the 6-year calculation). It is important to note that the compound annual growth rate depends only on the values for the first and last periods, so a series of values that fluctuates wildly during the period between the beginning and ending time periods would have the same compound annual growth rate as one which has a relatively steady climb. A slight modification of the method illustrated here would enable you to calculate the compound annual decline of a series of numbers that was going down—although they tend to be used less since investors are less interested in investments that are declining in value or in businesses with decreasing revenues or profits.

Calculations with Logarithms

Logarithms are seldom used for basic calculations now, but some familiarity with how they can be used is still valuable. Exhibit 7-2 shows three examples of calculations with logarithms. First we multiply two numbers by adding their logarithms. Since Mathcad can produce logarithms, and antilogarithms, on request, it is not necessary to use logarithm tables or to interpolate values to produce these examples. The second and third examples show how we raise a number to a power by multiplying its logarithm by the power. Of course, we could also divide by subtracting logarithms or find a root by dividing a logarithm by a power—but you get the idea.

Mutual fund A sold for $10 at the end of 1989 and for $35 at the end of 1994
What is the compound annual growth rate over that span?

Divide the ending price by the beginning price

$$r := \frac{35}{10} \qquad r = 3.5$$

The span of years is 5

Raise 3.5 to the 1/5 power

$3.5^{\frac{1}{5}} = 1.285$ So the compound annual growth factor is 1.285
                                In other words, the compound annual growth is 28.5%

Exhibit 7-1

Compounding

If you put some money in a bank account at the beginning of a year and the account pays 10 percent interest per year, compounded once a year, the bank will add your 10 percent interest to the account at the end of the year. Exhibit 7-3 shows how to calculate the results for accounts that are compounded more frequently: semiannually, quarterly, monthly, daily, hourly, and continuously.

Insects

A few years ago, a scientist was quoted in a major magazine as saying that there are 10^{18} insects in the world and that the total weight of all these insects was more than the weight of all the human beings in the world. For purposes of this example, we assume that there are about 5 billion people in the world and that the average person weighs about 70 kg (70,000 g). We further assume that the average insect weighs 2.5 mg. Then, assuming that our estimates are reasonably correct, our calculations in Exhibit 7-4 indicate that the scientist is right. In fact, the insects weigh about 11 times as much as the people.

Photons

In Exhibit 7-5, we calculate the number of photons of light emitted by a typical 100-W lightbulb. The energy of a single photon is determined by the frequency of the light. Having computed the energy of a photon, we go on to compute how many photons are being produced.

Multiply two numbers (2*3) by adding their logarithms

$\log(2) = 0.301029995663981$

$\log(3) = 0.477121254719662$

$0.301029995663981 + 0.477121254719662 = 0.778151250383643$

This is the logarithm of the answer

$10^{0.778151250383643} = 6$

This is the answer to the above multiplication problem (2*3=6)

Here we raise the number 5 to the power of 2, so we multiply its log by 2

$\log(5) = 0.698970004336019$

$a := \log(5)$

$2 \cdot a = 1.397940008672038$

$10^{1.397940008672038} = 25$

So we have demonstrated that 5 squared (i.e. raised to the second power) is 25

$c := \log(e)$ $c = 0.434294481903252$

We want to raise e to the i * pi power, so we multiply its logarithm by i and by pi

$c \cdot \sqrt{-1} \cdot \pi = 1.364376353841841i$

$10^{1.364376353841841i} = -1$

So we have demonstrated with logarithms the famous result that e raised to the power of i * pi is equal to -1

Exhibit 7-2

Annual Compounding	$1000 \cdot 1.1 = 1100.000000$
Semiannual	$1000 \cdot 1.05^2 = 1102.500000$
Quarterly	$1000 \cdot 1.025^4 = 1103.812891$
Monthly	$1000 \cdot \left(1 + \dfrac{.1}{12}\right)^{12} = 1104.713067$
Daily	$1000 \cdot \left(1 + \dfrac{.1}{365}\right)^{365} = 1105.155782$
Hourly	$1000 \cdot \left(1 + \dfrac{.1}{24 \cdot 365}\right)^{24 \cdot 365} = 1105.170287$
Continuously	$1000 \cdot e^{.1} = 1105.170918$

Exhibit 7-3

Assume that there are:

$1 \cdot 10^{18}$ insects in the world

Assume the average insect weighs 2.5 milligrams

$1.10 \cdot 10^{18} \cdot 2.5 \cdot 10^{-3} = 2.75 \cdot 10^{15}$ grams

Assume there are 5*10^9 people in the world
and assume an average of 50 kilograms (50,000 grams)
each, then the people weigh

$5 \cdot 10^9 \cdot 50000 = 2.5 \cdot 10^{14}$

So if these estimates are roughly correct, the bugs outweigh the people
by about a factor of:

$$\frac{2.75 \cdot 10^{15}}{2.5 \cdot 10^{14}} = 11$$

Exhibit 7-4

A 100 Watt lightbulb emits light with an average wavelength of $5.75 \cdot 10^{-7}$ meters.

How many photons does the bulb emit in one second?

The frequency of the light is equal to the distance light travels in one second c, divided by the wavelength of the light λ.

$$c := 3 \cdot 10^8 \cdot \frac{m}{sec} \qquad\qquad \lambda := 5.75 \cdot 10^{-7} \cdot m$$

$$f := \frac{c}{\lambda} \qquad\qquad f = 5.217 \cdot 10^{14} \ \bullet Hz$$

The energy of each photon is $h \cdot f$, where h is Planck's Constant

$$h := 6.63 \cdot 10^{-34} \cdot joule \cdot sec$$

$$E := h \cdot f \qquad\qquad E = 0 \bullet joule$$

Note that you must remember to reset the "zero tolerance" on the "numerical format" dialog box to a large value. Otherwise, Mathcad will assume you want the very small values in this problem rounded to zero.

Since this is a 100 Watt bulb, and 100 Watts = 100 joules/second, the number of photons emitted per second is

100 joules/second divided by the energy of a photon, or

$$\frac{100 \cdot \dfrac{joule}{sec}}{E} = 2.891 \cdot 10^{20} \ \bullet s^{-1}$$

So $2.981 \cdot 10^{20}$ photons are emitted each second

Exhibit 7-5

$x := \sqrt{2}$

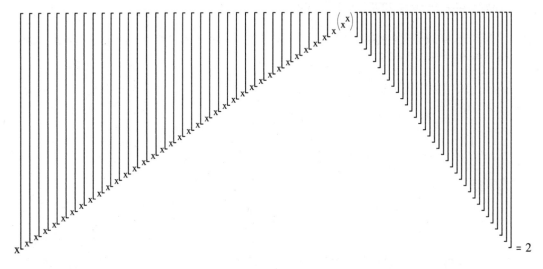

Exhibit 7-6

$$\sum_{n=1}^{99999999999999999999999999999999} 2 \cdot n$$

9999999999999999999999999999999990000000000000000000000000000000000000

Exhibit 7-7

$$\frac{76!}{32!}$$

71656166325392602998738516480418250097156776527222375301423038464000000000000

Exhibit 7-8

Powers of *x*

In Exhibit 7-6 we calculate a cascade of powers of x as a demonstration of the power of Mathcad to compute something that would otherwise be difficult to evaluate. You may find it interesting to play around with the value of x in this example. You will find that as x grows much larger than the square root of 2, the value quickly overflows. In fact, you will find that the best you can do is $x = e^{1/e}$, which gives the limit e.

Full-Precision Calculations

Full-precision calculations with the symbolic processor

In addition to working with symbolic expressions, the symbolic processor is a powerful tool for computing full-precision numerical answers to many mathematical expressions. In Exhibit 7-7 we see an impressive demonstration of this capability in a summation problem.

Full-precision factorial calculation

In Exhibit 7-8 we see a full-precision calculation involving division of large factorials.

Factoring

Prime-number factoring

Mathcad's "Factor Expression" on the Symbolics menu can also be used to factor large numbers as we see in Exhibit 7-9.

One-hundredth power of 3

In Exhibit 7-10, the one-hundredth power of 3 is calculated to full precision and then factored to produce the original expression, using the symbolic engine.

Factoring and expanding

The symbolic engine can expand and factor mathematical expressions as well as numbers. This is shown in Exhibit 7-11.

Power Series Expansion around a Specified Point

This example (Exhibit 7-12) is from a QuickSheet. You specify a function, a point around which it is to be expanded, and the highest-order term to be included in the expansion, and the QuickSheet will have Mathcad generate the appropriate power series expansion.

Catenary

A *catenary* is a curve made by freely suspending a cord of uniform weight between two points. Exhibit 7-13 gives an example and includes a graph.

Factor example

23754

$(2)\cdot(3)\cdot(37)\cdot(107)$

Second example

1024

$(2)^{10}$

Third example

234098234985023498987923 47

$(3)\cdot(26641802587)\cdot(32447507)\cdot(9026761)$

Exhibit 7-9

3^{100}

515377520732011331036461129765621272702107522001

You could also use the "Evaluate Symbolically" arrow from the "Evaluation and Boolean Palette" followed by "Enter."

$3^{100} \rightarrow$ 515377520732011331036461129765621272702107522001

Or to do the reverse, select "Factor Expression" from the symbolic menu

515377520732011331036461129765621272702107522001

$(3)^{100}$

Exhibit 7-10

Using Expand Expression from the Symbolics menu.

$$(1 + x)^{11}$$

$$x^{11} + 11 \cdot x^{10} + 55 \cdot x^9 + 165 \cdot x^8 + 330 \cdot x^7 + 462 \cdot x^6 + 462 \cdot x^5 + 330 \cdot x^4 + 165 \cdot x^3 + 55 \cdot x^2 + 11 \cdot x + 1$$

Using Factor Expression from the Symbolics menu.

$$x^{11} + 11 \cdot x^{10} + 55 \cdot x^9 + 165 \cdot x^8 + 330 \cdot x^7 + 462 \cdot x^6 + 462 \cdot x^5 + 330 \cdot x^4 + 165 \cdot x^3 + 55 \cdot x^2 + 11 \cdot x + 1$$

$$(x + 1)^{11}$$

Exhibit 7-11

From Mathcad QuickSheets

Expanding a Function in a Power Series

Uses live symbolic computation to expand a function
in a power series, using **series** keyword.

Enter a function f(x):

$$f(x) := \exp(x)$$

**Point around which to expand, and
order of highest order term:**

Note: series is a symbolic keyword.

Series expansion:

$$f(x) \text{ series}, x = 0, 7 \; \rightarrow \; 1 + x + \frac{1}{2} \cdot x^2 + \frac{1}{6} \cdot x^3 + \frac{1}{24} \cdot x^4 + \frac{1}{120} \cdot x^5 + \frac{1}{720} \cdot x^6$$

Exhibit 7-12

A catenary is a curve made by freely suspending a cord of uniform weight between two points.

ORIGIN := -100

x := -100.. 100

a := 10

a is the distance of the lowest hanging point of the cord above the ground.

y is the distance of any point on the cord above the ground.

$$y_x := \frac{a}{2} \cdot e^{\frac{x/5}{a}} + e^{\left[-\left(\frac{x/5}{a}\right)\right]}$$

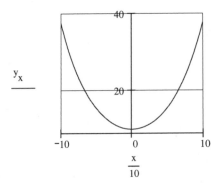

Exhibit 7-13

Three-Dimensional Graph

Mathcad's QuickSheet (Exhibit 7-14) gives the graph of a sphere. You can change the graph by redefining the value of N, which represents the number of vertical separations.

Stefan–Boltzmann Law

A version of the Stefan–Boltzmann law is demonstrated in Exhibit 7-15. The energy density of blackbody radiation is shown to increase rapidly with temperature because of the Stefan–Boltzmann fourth-power relationship. A second version of this important relationship is demonstrated in Exhibit 7-16.

From Mathcad's QuickSheets

Plotting a Sphere

Creates a surface plot of a unit sphere.

Enter the number of
vertical separations: $N := 25$

$$i := 0 .. N \qquad \phi_i := i \cdot \frac{\pi}{N} \qquad j := 0 .. N \qquad \theta_j := j \cdot 2 \cdot \frac{\pi}{N}$$

$$X_{i,j} := \sin(\phi_i) \cdot \cos(\theta_j) \qquad\qquad Y_{i,j} := \sin(\phi_i) \cdot \sin(\theta_j)$$

$$Z_{i,j} := \cos(\phi_i)$$

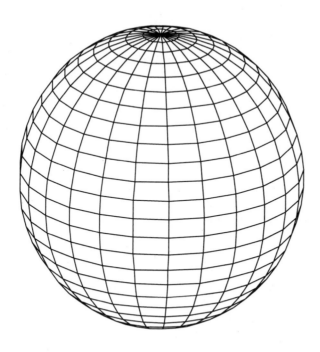

X, Y, Z

Exhibit 7-14

The Stefan-Boltzmann Law shows that the energy density of black body radiation is proportional to the fourth power of the temperature.

We want to measure energy density in electron volts per liter, so we must define electron volts in terms of Mathcad's built-in units.

$$ev := 1.60219 \cdot 10^{-12} \cdot erg$$

$$k := 4.72 \cdot \frac{ev}{liter \cdot K^4}$$ k is the proportionality constant between temperature and energy density

$$f(t) := k \cdot t^4$$ Here we define a function that produces the energy density given the temperature.

$$f(1 \cdot K) = 4.72 \cdot \frac{ev}{liter}$$ Here we see the rapid rise in energy density with temperature.

$$f(10 \cdot K) = 4.72 \cdot 10^4 \cdot \frac{ev}{liter}$$

$$f(100 \cdot K) = 4.72 \cdot 10^8 \cdot \frac{ev}{liter}$$

$$f(1000 \cdot K) = 4.72 \cdot 10^{12} \cdot \frac{ev}{liter}$$

$$f(3 \cdot K) = 382.32 \cdot \frac{ev}{liter}$$ Density for 3 degree background radiation from the Big Bang.

$$f(3000 \cdot K) = 3.823 \cdot 10^{14} \cdot \frac{ev}{liter}$$ Density for 1000 times the temperature of 3 degree radiation.

Exhibit 7-15

Here we derive the Stefan-Boltzmann Law from some basic constants including The Boltzmann's Constant, Planck's Constant and the Speed of Light.

$$T := 1 \cdot K$$

$$k := 1.38 \cdot 10^{-16} \cdot \frac{erg}{K}$$

$$c := 299729 \cdot 10^5 \cdot \frac{cm}{sec}$$

$$h := 6.625 \cdot 10^{-27} \cdot erg \cdot sec$$

$$u := \left[\frac{8 \cdot \pi^5 \cdot (k \cdot T)^4}{15 \cdot (h \cdot c)^3} \right]$$

$$u = 0 \cdot kg \cdot m^{-1} \cdot s^{-2}$$

$$u = 7.56 \cdot 10^{-15} \cdot \frac{erg}{cm^3}$$

Exhibit 7-16

Speed of Molecules

It is well known that molecules in a hot gas move more rapidly than do those in a cool gas. In Exhibit 7-17 we calculate the actual speed of oxygen molecules in a room at 300 K.

Engineering Studies for Engine Design

Two types of motion for an engine (crank and rod motion and simple harmonic motion) are compared in an engineering study for the design of an engine. The vertical piston motion (lift), the velocity (derivative of lift), the acceleration (derivative of velocity) as well as the jerk (derivative of acceleration) are shown in Exhibit 7-18.

Doppler Shift

In Exhibit 7-19, we calculate the value of the Doppler shift in the wavelength of a line of the spectrum from a distant group of galaxies. The Doppler shift arises from the motion of the distant galaxies away from our galaxy.

Temperature is the average random kinetic energy of molecules.

The average kinetic energy is $\frac{3}{2}\cdot k\cdot t$

where k is Boltzmann's Constant

$$k := 1.3806\cdot 10^{-16}\cdot\frac{erg}{K}$$

A Dalton is the molecular mass unit.

We need to define the Dalton for Mathcad in terms of Mathcad's built-in units.

$$Dl := \frac{1}{6.02\cdot 10^{26}}\cdot kg$$

What is the average speed of an oxygen molecule in a room
at a temperature of 300 K?

The mass of an oxygen molecule is 32 Dl

$$m := 32\cdot Dl$$

$$T := 300\cdot K$$

The formula for the speed is:

$$s := \sqrt{\frac{3\cdot k\cdot T}{m}}$$

$s = 483.5\bullet m\bullet s^{-1}$ Is the average speed of the oxygen molecules.

Exhibit 7-17 Speed of molecules.

Crank and Rod Piston motion curves

c := 120   Connecting rod length (in millimeters)

r := 34   One half the stroke (in millimeters)

$$y(t) := r + c - r \cdot \cos(t) - \sqrt{c^2 - (r \cdot \sin(t))^2}$$

$$v(t) := \frac{d}{dt} y(t) \qquad a(t) := \frac{d^2}{dt^2} y(t) \qquad j(t) := \frac{d^3}{dt^3} y(t)$$

$$m(t) := r - r \cdot \cos(t)$$

$$n(t) := \frac{d}{dt} m(t) \qquad o(t) := \frac{d^2}{dt^2} m(t) \qquad p(t) := \frac{d^3}{dt^3} m(t)$$

$$t := 0, 2 \cdot \frac{\pi}{180} .. 2 \cdot \pi \qquad u := \frac{360}{2 \cdot \pi}$$

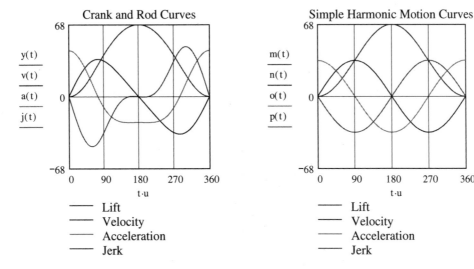

Crank and Rod Curves	Simple Harmonic Motion Curves

Lift
Velocity
Acceleration
Jerk

Exhibit 7-18 Engineering studies for engine design.

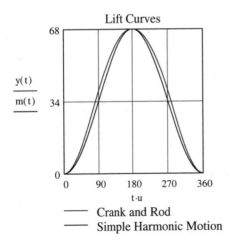

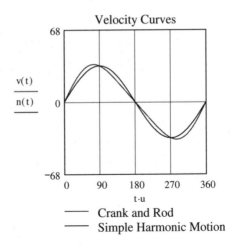

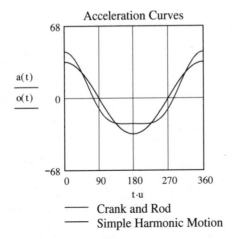

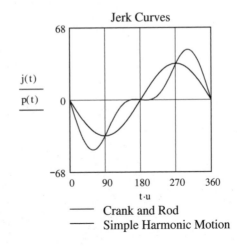

Exhibit 7-18 (*Continued*)

The galaxies of the Virgo cluster are moving away from our galaxy at a speed of about 1000 km per second. Therefore, the wavelength of any line in the spectrum of the Virgo cluster will be Doppler shifted from its normal value.

$s := 1000 \cdot \dfrac{km}{sec}$ Speed of the cluster away from our galaxy.

$c := 300000 \cdot \dfrac{km}{sec}$ Speed of light.

$R := 1 + \dfrac{s}{c}$ Ratio of the shifted wavelength to the normal wavelength.

$R = 1.0033$

Exhibit 7-19 Doppler shift.

Mass of Sunlight Striking the Earth

In Exhibit 7-20, we calculate the flux of energy from the sun striking the earth and, using Einstein's famous equation for the equivalence of energy and matter, we determine how much mass the sunlight striking the earth in one second represents.

Simple Harmonic Motion

The differential equation that represents simple harmonic motion that is damped, but free (i.e., with no external driving force), has solutions which fall into one of three types. These are shown in Exhibits 7-21 to 7-23.

Black–Scholes Option Pricing

An investor decides to generate some income by writing a call option. He uses the Black–Scholes option pricing model to determine the price. The calculation is illustrated in Exhibit 7-24.

First we compute the surface area of a sphere centered on the Earth, with the radius of the Earth's average distance from the Sun.

$R := 93000000 \cdot mi$

$S := 4 \cdot \pi \cdot R^2$

$S = 2.815 \cdot 10^{23} \cdot m^2$

Then we calculate the area of the disk of the Earth, to determine the fraction of the Sun's energy flux that is intercepted by the Earth.

$r := 3963 \cdot mi$

$a := \pi \cdot r^2$

$a = 1.278 \cdot 10^{14} \cdot m^2$

The energy output of the Sun is:

$E := 3 \cdot 10^{33} \cdot \dfrac{erg}{sec}$

$E = 3 \cdot 10^{26} \cdot s^{-1} \cdot joule$

$E = 3 \cdot 10^{26} \cdot \dfrac{joule}{sec}$

So the portion of the Sun's flux of energy, at the distance of Earth's orbit, that is intercepted by the Earth is approximately:

$\dfrac{a}{S} \cdot E = 1.362 \cdot 10^{17} \cdot kg \cdot m^2 \cdot s^{-3}$

$\dfrac{a}{S} \cdot E = 1.362 \cdot 10^{17} \cdot \dfrac{joule}{sec}$

The speed of light is:

$c := 300000 \cdot \dfrac{km}{sec}$

$m := \dfrac{\dfrac{a}{S} \cdot E}{c^2}$

$m = 1.513 \cdot kg \cdot s^{-1}$ So each second, the Earth is struck by about one and one half kilograms of sunlight.

Exhibit 7-20

The equation is $m \cdot \dfrac{d^2}{dt^2} y + c \cdot \dfrac{d}{dt} y + k \cdot y = 0$

$t := 0 \ldots 1000$

Let the constants have the values

$m := 20000$

$c := 7000$

$k := 300$

This is Case 1, because $c^2 - 4 \cdot k \cdot m > 0$

Case 1, $\quad c^2 - 4 \cdot k \cdot m > 0$ $\qquad\qquad c^2 - 4 \cdot k \cdot m = 2.5 \cdot 10^7$

This is overdamped motion. It behaves sluggishly and returns to zero slowly.

$r1 := \dfrac{-c + \sqrt{c^2 - 4 \cdot k \cdot m}}{2 \cdot m}$ $\qquad\qquad r2 := \dfrac{-c - \sqrt{c^2 - 4 \cdot k \cdot m}}{2 \cdot m}$

$r1 = -0.05$ $\qquad\qquad\qquad r2 = -0.3$

$a := 1 \qquad b := 2$

$y_t := a \cdot e^{r1 \cdot \frac{t}{100}} + b \cdot e^{r2 \cdot \frac{t}{100}}$

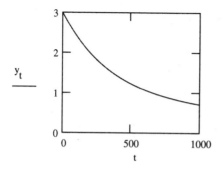

Exhibit 7-21

t := 0.. 1000

Let the constants have the values

m := 30000

c := 60000

k := 30000

Case 2, $c^2 - 4 \cdot k \cdot m = 0$ $c^2 - 4 \cdot k \cdot m = 0$

This is critically damped motion. It returns to equilibrium in the shortest time and without overshoot.

a := 1   b := 1.5

$$y_t := \left(a + b \cdot \frac{t}{100}\right) \cdot e^{-c \cdot \frac{\frac{t}{100}}{2 \cdot m}}$$

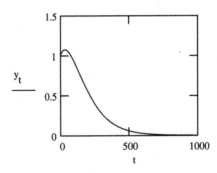

Exhibit 7-22

$t := 0 .. 1000$

Let the constants have the values

$m := 20000$

$c := 700$

$k := 1000$

Case 3, $c^2 - 4 \cdot k \cdot m < 0$ $c^2 - 4 \cdot k \cdot m = -7.951 \cdot 10^7$

This is underdamped motion. It overshoots and oscillates before reaching equilibrium.

$a := 5$

$b := 7$

$$\mu := \frac{\sqrt{4 \cdot k \cdot m - c^2}}{2 \cdot m}$$

$$y_t := e^{\frac{-c \cdot \frac{t}{10}}{2 \cdot m}} \cdot \left(a \cdot \cos\left(\mu \cdot \frac{t}{10} \right) + b \cdot \sin\left(\mu \cdot \frac{t}{10} \right) \right)$$

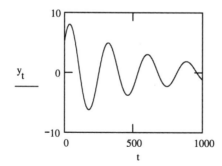

Exhibit 7-23

An investor wants to earn income by writing a call option on the stock of XYZ company, which is currently selling for $50 a share. He plans to write a 3-month call option with an exercise (or striking) price of $50. He wants to use the Black-Scholes formula to determine the appropriate premium to charge.

$S := 50$ the current price of XYZ stock.

$k := 50$ the exercise price of the call option.

$R := .09$ the continuously compounded risk-free investment rate.

$\sigma := \sqrt{.25}$ σ^2 is the variance of XYZ stock's return.

$T := \dfrac{3}{12}$ time (in fractions of a year) before the option matures.

$$D1 := \frac{\ln\left(\dfrac{S}{k}\right) + \left(R + .5 \cdot \sigma^2\right) \cdot T}{\sigma \cdot \sqrt{T}} \qquad D1 = 0.215$$

$$D2 := D1 - \sigma \cdot \sqrt{T} \qquad\qquad D2 = -0.035$$

$$N1 := \mathrm{cnorm}(D1) \qquad\qquad N1 = 0.585$$

$$N2 := \mathrm{cnorm}(D2) \qquad\qquad N2 = 0.486$$

Then the cost or premium of the call, C, is given by

$$C := S \cdot \mathrm{cnorm}(D1) - e^{-R \cdot T} \cdot k \cdot \mathrm{cnorm}(D2)$$

$C = 5.49$ So the investor should charge a $5.49 premium for the call option on XYZ stock.

Exhibit 7-24

Economic-Order Quantity

In Exhibit 7-25 we see an example from business management of the calculation of the optimum batch size for ordering supplies. The problem arises because a business can avoid shortages of inventory by ordering, and keeping in stock, large amounts of important supplies or components or products to be sold. But keeping large quantities in inventory has its own costs (in storage, security, wastage, etc.). Formulas such as this one are intended to minimize the overall cost by optimizing the amount purchased in each batch, and hence the frequency of replenishment of stocks.

If   d = annual demand for product
Q = size of batch
t = time interval between replenishment orders (in years)
     also, t=Q/d

p = cost price per item
     the stock-holding cost per item per year = i*p
     and the total annual stock-holding cost = i*p*Q/s

     the number of deliveries per year = d/Q

     if c = delivery cost per batch,

     the annual delivery cost = c*d/Q

Using differentiation, it can then be shown that the batch size which makes the total variable cost take its minimum value is for

$$Q = \sqrt{\frac{2 \cdot c \cdot d}{i \cdot p}}$$

So if

c := 10

d := 600

i := .2

p := 6

$$Q := \sqrt{\frac{2 \cdot c \cdot d}{i \cdot p}} \qquad Q = 100$$

Exhibit 7-25

8

Electronic Aids to Problem Solving

In the previous chapters we have seen examples of many ways to use Mathcad's mathematical power in solving a wide array of problems. In this chapter we introduce some of the electronic aids that are available to help you learn to use Mathcad and apply it to your own work. Many Mathcad solution procedures have already been developed for your use and can be readily adapted to your specific problem. Mathcad can save you from having to reinvent the wheel, so to speak, when you encounter a new type of problem. We also show how you can produce 3D plots of your data.

QuickSheets

QuickSheets are helpful for performing frequently used tasks. They range from simple operations to complex programs. They are in the Resource Center and can be accessed from either the "Help" menu or the "Resource Center" button, shown in Exhibit 8-1. Exhibit 8-2 is an example of a QuickSheet. The most useful feature of QuickSheets is that they can be temporarily modified with your data to obtain a quick answer to your own particular problem. Any modifications to the permanent QuickSheets are never saved, but Mathcad version 7.0 gives you four personal QuickSheets, which can be modified and saved. Last, any part of the QuickSheet can be copied onto your worksheet by simply selecting the desired selection and dragging it onto your worksheet. You can also use "copy" and "paste" from the QuickSheet toolbar to move the QuickSheet programs into your work.

Click on the special QuickSheet button shown in Exhibit 8-3 to get a reminder of how to use the sheet in your Mathcad document. There are well over 100 QuickSheet problems that can be readily incorporated into your work. To give you an idea of the range of topics QuickSheets can help you with, following is a list of QuickSheet contents from the Resource Center.

About QuickSheets

Personal QuickSheets

Arithmetic and Algebra

Exhibit 8-1 Resource
center button.

Business and Finance

Units

Vector and Matricies

Solving Equations

Graphing and Visualizations

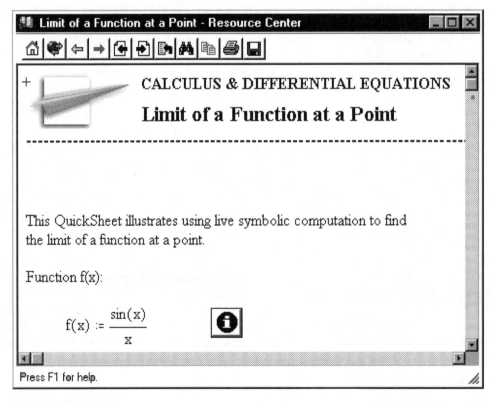

Exhibit 8-2 QuickSheet.

Exhibit 8-3 Special
QuickSheet button.

Calculus and Differential Equations

Data Analysis

Statistics

Components in Mathcad

Special Functions

Programming

Animations

Extra Math Symbols

You can see from the list of topics that many examples can be used to help you learn the use and functionality of Mathcad.

Polynomial curve fitting

As an example of how you can use QuickSheets, one of the authors uses the polynomial curve fitting QuickSheet regularly in his work. This is a powerful and easily used tool for analysis of your data. Most recently it was used in the calibration of a air-mass flowmeter. Once the data has been collected in a file (see Exhibit 8-4), the QuickSheet is selected. The name of the

.2342	-15.0
.3520	-12.0
.4393	-10.0
.4850	-9.0
.5329	-8.0
.5838	-7.0
.6386	-6.0
.6981	-5.0
.7625	-4.0
.8303	-3.0
.8997	-2.0
.9964	0.0
1.0932	2.0

Exhibit 8-4 Text from
file "MassFlow.Prn."

input file is changed and perhaps the order number of the polynomial. The page with the results is then printed. See Exhibit 8-5 for an example of the results.

Personal QuickSheets

Mathcad provides the ability for you to create your own QuickSheets. Your personal QuickSheets can be used to save useful Mathcad expressions, information, hypertext links, custom operators, and more. You can change personal QuickSheets temporarily or permanently. There are four such personal sheets provided for your use. They are named as follows:

My Formulas

My Programs

My Operators

My Hot Links

Although the pages have specific names, the names can be changed to suit your personal preferences, and any information can be saved on any page.

An example of a personal program is shown in Exhibit 8-6. This program is used to plot a port time area diagram for a port on a two-stroke engine.

Mathcad Treasury

Impressive as QuickSheets are as a resource to the Mathcad user, there is an even more comprehensive and powerful on-line help facility called the *Mathcad Treasury*. The extra-cost Treasury contains a true treasure-house (hundreds of files) of mathematical techniques that can be incorporated into your work or browsed and studied to enhance your knowledge of both mathematics and Mathcad. The following list gives a taste of what is available.

Arithmetic

Basic Functions

Linear Algebra and Matrix Arithmetic

Complex Numbers

Graphing and Plotting in the Plane

Plotting Multi-dimensional Data

Calculus

Probability

Statistics

Solving

Applied Mathematics

Symbolic Computation

Polynomial Regression

Uses Mathcad statistical functions for
polynomial regression of X-Y data.

Enter a matrix of X-Y data to be analyzed:

data := READPRN("MassFlow.prn")

Scroll right to see the matrix data used in this example.

data =

$X := data^{<0>}$ $Y := data^{<1>}$ $n := rows(data)$

**Enter degree of polynomial
to fit:** $k := 2$

Number of data points: $n =$

$z := regress(X, Y, k)$

Polynomial fitting function: $fit(x) := interp(z, X, Y, x)$

$coeffs := submatrix(z, 3, length(z) - 1, 0, 0)$

Coefficients: $(coeffs)^T =$

R²: $\dfrac{\sum(\overrightarrow{fit(X) - mean(Y)})^2}{\sum(\overrightarrow{Y - mean(Y)})^2} =$

Degrees of freedom: $n - k - 1 =$

Exhibit 8-5 Calibration of an air-mass flowmeter.

Plots

$i := 0 .. n - 1$ $j := 0 .. 49$

$$tx_j := min(X) + j \cdot \frac{max(X) - min(X)}{50}$$

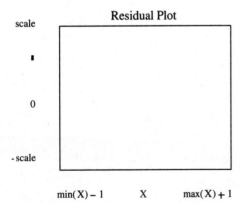

$$scale := max\left(\left| \overrightarrow{fit(X) - Y} \right|\right) \cdot 1.1$$

Residual Plot

scale

0

- scale

min(X) − 1   X   max(X) + 1

Exhibit 8-5 *(Continued)*

My Programs

$S := 50.6$ Stroke

$C := 110$ Connecting Rod Length

$W := 5$ Wrist Pin Offset Sign is CRITICAL!!! (Positive in clockwise rotation when offset is to the right)

(Zero Degrees is at, or near, Top Dead Center depending on wrist pin offset)

$POA := 115$ Approximate angle where port begins to open (in degrees)

$FOA := 180$ Approximate angle where port is fully opened (in degrees)

$HS := \dfrac{S}{2}$ Half the Stroke $HS = 25.3$

$B := C + HS$ Top of the Block $B = 135.3$

$H(ang) := \cos(ang) \cdot HS + \sqrt{C^2 - (\sin(ang) \cdot HS - W)^2}$ Piston position as function of crank rotation

$TDC := H(0)$ Height of Piston at Top Dead Center (should equal B if W = 0) $TDC = 135.186$

$HPO := H(POA \cdot deg)$ Height at Port Open $HPO = 97.837$

$PO := TDC - HPO$ Port Open Dimension $PO = 37.35$

$HPFO := H(FOA \cdot deg)$ Height at Port Full Open $HPFO = 84.586$

$PH := HPO - HPFO$ Port Height Dimension $PH = 13.25$

$YPO := \begin{bmatrix} 0 \\ HS \cdot \cos(POA) \\ H(POA) \end{bmatrix}$ $XPO := \begin{bmatrix} 0 \\ HS \cdot \sin(POA) \\ W \end{bmatrix}$ $i := 0 .. 2$

$ang := 0, 0.1 .. 2 \cdot \pi$

$Xmax := \text{ceil}(HS + C)$

$Ymax := Xmax$

Exhibit 8-6

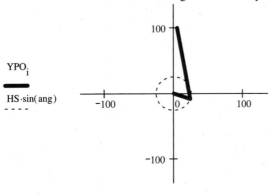

Crank And Connecting Rod at Port Open

$$A(P) := 2 \cdot \mathrm{atan}\left[\frac{2 \cdot HS \cdot W + \sqrt{2 \cdot \left(HS^2 + C^2\right) \cdot \left(W^2 + P^2\right) - \left(HS^2 - C^2\right)^2 - \left(W^2 + P^2\right)^2}}{W^2 + (HS + P)^2 - C^2}\right]$$ Angle from Piston Position

$\dfrac{A(98)}{deg} = 114.559$ Actual Port Open Angle obtained from the piston position

Table showing the Crank Angle that produces the maximum piston lift
Only important if there is wrist pin offset in the piston.

$Ang := -.02, -.015 .. .05$ range variable from before TDC to after TDC

Crank Angle Ang	Piston Height H(Ang)
-0.02	135.157
-0.015	135.166
-0.01	135.173
$-5 \cdot 10^{-3}$	135.18
0	135.186
$5 \cdot 10^{-3}$	135.192
0.01	135.196
0.015	135.2
0.02	135.203
0.025	135.205
0.03	135.207
0.035	135.208
0.04	135.207
0.045	135.207
0.05	135.205

Exhibit 8-6 *(Continued)*

Common Functions

Common Curves and Surfaces

Advanced Data Analysis

Wavelets

Matrix Algebra

Differential Equations

Extended Symbolic Capabilities

Live Symbolics and Optimization

Advanced Programming Techniques

Animation

Index

Reference Tables

References tables are included in the Resource Center. The following topics are contained in the Reference Tables.

Basic Science References

Calculus Reference Formulas

Properties of Metals

Properties of Liquids

Properties of Solids

Properties of Gases

Earth Science

Geometry Formulas

Mechanics and Electromagnetics

Electronic Books

Electronic Books are another extra-cost valuable source of information for your problem solving. New electronic books on a variety of topics are always being published by MathSoft. Here is a list of recently available electronic books.

Extension Packs

Mathcad Selections from Numerical Recipes

Signal Processing

Image Processing

Shaum's Electronic Tutors (available from McGraw-Hill or MathSoft)

   Electric Circuits

   Feedback and Control Systems

   Electromagnetics

   Thermodynamics for Engineers

   Fluid Mechanics and Hydraulics

   College Physics

   College Algebra

   Statistics

   Engineering Mechanics

   Mechanical Vibrations

Mathcad Education Library

   Calculus

   Introduction to Ordinary Differential Equations

General Interest Electronic Books

   Astronomical Formulas

   Personal Finance

   The McGraw-Hill Financial Analyst

Mathcad Electronic Books

   Electrical Engineering Library

   Mechanical Engineering Library

Gieck's Engineering Formulas

Queuing Theory

A continually updated online source of additional Mathcad .mcd and .html resources is maintained by MathSoft Inc. at its site on the World Wide Web. You can access this site easily by choosing the Web Library from the Resource Center. This will automatically go through your Internet service provider to http://www.mathsoft.com. See Chap. 9 of this book for more on using Internet resources with Mathcad.

Plotting in Three Dimensions

There are many types of three-dimensional plots available to you in Mathcad including surface plots, contour plots, 3D bar charts, 3D scatter plots, and vector field plots. In Exhibit 8-7 we see an example of how you can easily create a variety of types of 3D plots from the same set

$$
A := \begin{bmatrix} 1 & 11 & 21 & 31 & 41 & 51 \\ 2 & 12 & 22 & 32 & 42 & 52 \\ 3 & 13 & 23 & 33 & 43 & 53 \\ 4 & 14 & 24 & 34 & 44 & 4 \\ 5 & 15 & 25 & 35 & 45 & 3 \\ 16 & 8 & 5 & 4 & 3 & 2 \end{bmatrix}
$$

Surface Plot

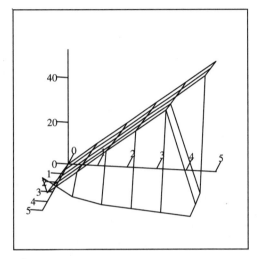

A

Contour Plot

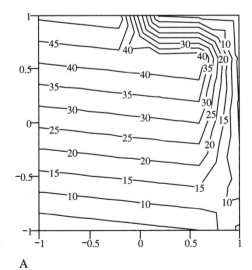

A

Exhibit 8-7

3-D Scatter Plot

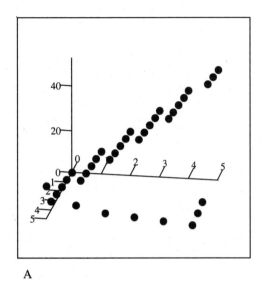

A

Vector Field Plot

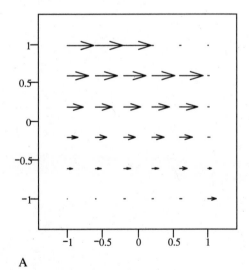

A

Exhibit 8-7 (*Continued*)

3-D Bar Chart

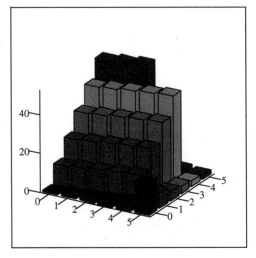

This chart has been rotated using "3D Plot Format" on the "Graphics" menu to 300 degrees in order to better show the bars.

A

Exhibit 8-7 (*Continued*)

---

of data. The key step is to enter your data into a matrix. In our example, we called the matrix "A." Then you can create a variety of plots by just selecting the appropriate type of plot from the "Graphics" menu and entering the name of the matrix containing the data into the placeholder. There is a lot you can do to fine-tune, label, and format your plots, but you can get started with 3D plotting by just doing what we have demonstrated here.

Putting Mathcad to Work with Other Programs

This chapter is about Mathcad's ability to work with other programs. Mathcad has long had this capability, but several enhancements and new features have been added in version 7. These add to your experience of using Mathcad by allowing you to do things that Mathcad alone could not do.

New to Mathcad 7 is support for OLE 2, which allows you to drag data from Mathcad to other programs that support OLE 2 (or vice versa) and to edit the resulting mathematics without returning to Mathcad. For example, a section of text and equations prepared in Mathcad could be moved into a Word (or Excel) document by simply dragging and dropping. You simply arrange the document windows on your computer screen so that you can select the part to be moved, then drag it out of your Mathcad document and into the window for the Word document. If you then want to change something in the section that came from Mathcad, just click on it and edit it as you would do in Mathcad. Alternatively, you could follow the same process if you wanted to move data from Word into Mathcad—or among any other OLE 2–compliant programs.

A new program called *MathConnex* that is supplied with Mathcad 7 Professional allows you to connect Mathcad with other programs (for example, Excel or MATLAB) to easily create and run a computational system. You can also create a presentation of your work that will display the desired level of detail. MathConnex allows you to drag icons representing different programs to the MathConnex document window and connect the appropriate inputs and outputs by simply drawing lines to show how they should be connected. This is discussed in the section on using MathConnex.

As useful as MathConnex and OLE 2 are in getting results from Mathcad and other programs, there are times when you just need to transfer data between programs. The remainder of this chapter describes in detail how to copy (or move) actual data and text between Mathcad and various Microsoft products when it is undesirable or just unnecessary to use OLE. When you move data as described here, the documents will not be linked to each other; therefore changes in one document will not affect the other document. These explanations will help you know what is possible and will likely save you valuable hours of experimentation and possible frustration when trying to share results between Mathcad and other programs. There is a section for each of the following products:

Microsoft Word

Microsoft Excel

Microsoft Access

Microsoft Paint

Microsoft Notepad

Microsoft Word

Numbers from Microsoft Word

Numerical data can be transferred directly from Microsoft Word into Mathcad in several different ways. A single column of numbers can simply be copied into the clipboard and then pasted into an input placeholder on a Mathcad worksheet. See Exhibits 9-1 and 9-2 for the before and after examples. Be careful; for this to work properly, there cannot be any extraneous characters such as blanks, commas, or extra carriage returns included in the data being copied. The message "This expression is incomplete. You must fill in the placeholders" may appear when extraneous characters exist. Eliminate the extraneous characters by selecting the data

$$v := \begin{bmatrix} 11 \\ 22 \\ 33 \\ 44 \\ 55 \end{bmatrix} \qquad \text{<== Copy into here}$$

11
22
33
44
55

$$v = \begin{bmatrix} 11 \\ 22 \\ 33 \\ 44 \\ 55 \end{bmatrix}$$

End of list . . .

Exhibit 9-1 Single column of numbers that will be copied into an input placeholder on a Mathcad worksheet (Exhibit 9-2.mcd). (Be careful not to get any extraneous characters.)

Exhibit 9-2 Single column of numbers copied into an input placeholder using Microsoft Word (from Exhibit 9-1.doc).

11	66
22	77
33	88
44	99
55	111

End of list . . .

Exhibit 9-3 Multiple column of numbers
that will be copied into an input
placeholder on a Mathcad worksheet
(Exhibit 9-4.mcd). (Be careful not
to get any extraneous characters.)

and attempting to copy it again. Turning on the paragraph symbols in Word helps show these extraneous characters.

Multiple columns of data can be copied under two different scenarios. If the data contains only the numbers separated by tabs (or commas) and no extraneous characters such as blanks or extra carriage returns, then they can be copied and then pasted into an input placeholder on the worksheet. See Exhibits 9-3 and 9-4 for the before and after multiple-column examples.

$$v := \begin{bmatrix} 11 & 66 \\ 22 & 77 \\ 33 & 88 \\ 44 & 99 \\ 55 & 111 \end{bmatrix} \qquad \text{<== Copy into here}$$

$$v = \begin{bmatrix} 11 & 66 \\ 22 & 77 \\ 33 & 88 \\ 44 & 99 \\ 55 & 111 \end{bmatrix}$$

Exhibit 9-4 Multiple column of numbers
copied into an input placeholder using
Microsoft Word (from Exhibit 9-3.doc).

10	11
12	13
14	15
16	17
18	19

End of list . . .

Exhibit 9-5 Table of numbers that will be copied into an input placeholder on a Mathcad worksheet (Exhibit 9-6.mcd). (Be careful not to get any extraneous characters.)

An array can also be copied if it exists in a Word *table*. See Exhibits 9-5 and 9-6 for the before and after table examples. Select the data within the table to be copied and then paste it onto the worksheet. If this method does not work for any reason, then the Word document can be saved as a "text only" file and the data can be transferred as described below for Microsoft Notepad.

$$v := \begin{bmatrix} 10 & 11 \\ 12 & 13 \\ 14 & 15 \\ 16 & 17 \\ 18 & 19 \end{bmatrix} \quad \text{<== Copy into here}$$

$$v := \begin{bmatrix} 171 \\ 272 \\ 373 \\ 474 \\ 575 \end{bmatrix}$$

$$v = \begin{bmatrix} 10 & 11 \\ 12 & 13 \\ 14 & 15 \\ 16 & 17 \\ 18 & 19 \end{bmatrix}$$

$$v = \begin{bmatrix} 171 \\ 272 \\ 373 \\ 474 \\ 575 \end{bmatrix} \quad \text{<== Copy from here}$$

Exhibit 9-6 Multiple column of numbers copied from a table into an input placeholder using Microsoft Word (from Exhibit 9-5.doc).

Exhibit 9-7 Single column of numbers copied from an output placeholder using Microsoft Word (into Exhibit 9-8.doc).

171
272
373
474
575

End of list . . .

Exhibit 9-8 Single column of numbers
  that were copied from an output placeholder on a Mathcad worksheet (Exhibit 9-9.mcd).

Numbers into Microsoft Word

Numerical data can be transferred into Microsoft Word by copying an array from an output placeholder into the clipboard and then pasting it into your Word document. The pasted text will contain only the numbers and tabs as separators for the columns. See Exhibits 9-7 and 9-8 for the before and after examples of a single column. See Exhibits 9-9 and 9-10 for the before and after examples of multiple columns.

$$v := \begin{bmatrix} 111 & 666 \\ 222 & 777 \\ 333 & 898 \\ 444 & 999 \\ 555 & 1101 \end{bmatrix}$$

$$v = \begin{bmatrix} 111 & 666 \\ 222 & 777 \\ 333 & 898 \\ 444 & 999 \\ 555 & 1.101 \cdot 10^3 \end{bmatrix} \qquad \text{<== Copy from here}$$

Exhibit 9-9 Multiple column of numbers
  copied from an output placeholder using
  Microsoft Word (into Exhibit 9-10.doc).

```
111   666
222   777
333   898
444   999
555   1.101E+3
```

This is line 1 Exh. 9-11
This is line 2 Exh. 9-11
This is line 3 Exh. 9-11
This is line 4 Exh. 9-11
This is line 5 Exh. 9-11

End of list . . .

Exhibit 9-10 Multiple column of numbers that were copied from an output placeholder on a Mathcad worksheet (Exhibit 9-9.mcd).

Exhibit 9-11 Text to COPY into a Mathcad worksheet (Exhibit 9-12.mcd normal and Exhibit 9-13.mcd using "Paste Special" and "unformatted text").

Text from Microsoft Word

Text can be transferred from Microsoft Word by copying it into the clipboard and then pasting it onto a Mathcad worksheet. See Exhibits 9-11 and 9-12 for the before and after text paragraph examples. The text will be deposited as a paragraph onto the worksheet. A paragraph differs from the normal text region because it is always a full page in width and cannot occupy the same space as any other object. To save text as a text region, select "Paste Special" from the edit menu, then select "unformatted text" from the list of options, then finally click on "OK." The pasted text will then be converted to a text region. Pasting the text into an existing text region also produces the desired result. See Exhibits 9-11 and 9-13 for the before and after text region examples.

Text into Microsoft Word

Text from a text region in a worksheet can be copied onto the clipboard and then pasted onto a Word document. The clipboard contains the rich text format of the text that is copied, and Mathcad uses this formatting information when it pastes the text onto the document. See

This is line 1 Exh. 9-11
This is line 2 Exh. 9-11
This is line 3 Exh. 9-11
This is line 4 Exh. 9-11
This is line 5 Exh. 9-11

This is line 1 Exh. 9-11
This is line 2 Exh. 9-11
This is line 3 Exh. 9-11
This is line 4 Exh. 9-11
This is line 5 Exh. 9-11

Exhibit 9-12 Text copied from Microsoft Word (Exhibit 9-11.doc) pastes as a paragraph (not a text region).

Exhibit 9-13 Text copied from Microsoft Word (Exhibit 9-11.doc) pasted "Paste Special" as a text region.

Line 1 - Exh. 9-14
Line 2 - Exh. 9-14
Line 3 - Exh. 9-14
Line 4 - Exh. 9-14
Line 5 - Exh. 9-14

Exhibit 9-14 Text copied to Microsoft Word (Exhibit 9-15.doc).

Line 1 - Exh. 9-14
Line 2 - Exh. 9-14
Line 3 - Exh. 9-14
Line 4 - Exh. 9-14
Line 5 - Exh. 9-14

Exhibit 9-15 This text was pasted directly from a Mathcad worksheet (Exhibit 9-14.mcd).

Exhibits 9-14 and 9-15 for the before and after text examples. Also, the worksheet as a whole can be exported as a rich-text-formatted (RTF) document that can be read directly into Microsoft Word. This will convert all the data, text, and graphics from the original worksheet. See Exhibits 9-16 and 9-17 for the before and after RTF examples.

Graphics from Microsoft Word

Word has two types of graphics. The first is a "Picture Object" which is drawn directly onto the document. The second type is just a "Picture" and results when a picture is created using the picture button. Picture Objects from Microsoft Word can be copied onto the clipboard and then pasted onto a worksheet using "Paste Special" and selecting "Word Picture Object" and pressing "OK." See Exhibits 9-18 and 9-19 for the before and after Picture Object examples. Pictures from Microsoft Word that are copied onto the clipboard and then pasted onto a worksheet using "Paste Special" and selecting Word Picture and pressing OK will be an OLE object.

Graphics into Microsoft Word

Pictures from a worksheet can be copied onto the clipboard and then pasted onto a Word document. Paste the picture using Paste Special and then either the Picture or "Bitmap" option before clicking OK. See Exhibits 9-20 and 9-21 for the before and after Picture examples.

Notice that the object has been changed by this process—it is now a Picture on the document (not a Picture Object as it would be if it had been originally transferred from Word).

Line 1 - Exh. 9-16
Line 2 - Exh. 9-16
Line 3 - Exh. 9-16
Line 4 - Exh. 9-16
Line 5 - Exh. 9-16

Exhibit 9-16 Text exported to Microsoft Word (Exhibit 9-17.rtf).

Line 1 - Exh. 9-16
Line 2 - Exh. 9-16
Line 3 - Exh. 9-16
Line 4 - Exh. 9-16
Line 5 - Exh. 9-16

Exhibit 9-17 Text exported to Microsoft Word (Exhibit 9-17.rtf).

Exhibit 9-18 A "Picture Object" copied to a Mathcad worksheet (Exhibit 9-19.mcd).

Exhibit 9-19 A "Picture Object" copied from Microsoft Word (Exhibit 9-18.doc) using "Paste Special."

Microsoft Excel

Numbers from Microsoft Excel

Numerical data can be transferred easily from Microsoft Excel into Mathcad. One or more columns of numbers can simply be copied into the clipboard and then pasted into an input placeholder on a Mathcad worksheet. See Exhibits 9-22 and 9-23 for the before and after examples. There can be no extraneous characters such as blanks, commas, or letters included in the data being copied. The message "This expression is incomplete. You must fill in the placeholders" may appear when extraneous characters exist. Eliminate the extraneous characters by selecting the data and attempt to copy it again.

$i := 1 .. 100$

$$s_i := \sin\left(\frac{i}{20}\right)^2$$

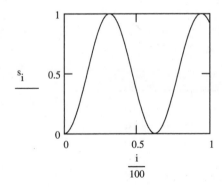

<== copied to Microsoft Word (Exhibit 09-21.doc)

Exhibit 9-20

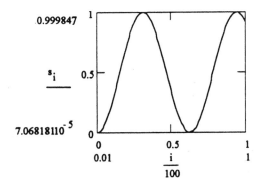

11	66
22	77
33	88
44	99
55	111

Exhibit 9-21 This picture was pasted special using "Picture" option on a graph from a Mathcad worksheet (Exhibit 9-20.mcd).

Exhibit 9-22 Multiple column of numbers that will be copied into an input placeholder on a Mathcad worksheet (Exhibit 9-21.mcd). (Be careful not to get any extraneous characters.)

Numbers into Microsoft Excel

Numerical data can be transferred into Microsoft Excel by copying an array and then pasting it into your Excel document. The pasted text will contain only the numbers and tabs as separators for the columns. See Exhibits 9-24 and 9-25 for the before and after examples of a copying columns.

$$v := \begin{bmatrix} 11 & 66 \\ 22 & 77 \\ 33 & 88 \\ 44 & 99 \\ 55 & 111 \end{bmatrix}$$ <== Pasted from an Excel spreadsheet

$$v = \begin{bmatrix} 11 & 66 \\ 22 & 77 \\ 33 & 88 \\ 44 & 99 \\ 55 & 111 \end{bmatrix}$$

Exhibit 9-23 These multiple columns of numbers were copied from Microsoft Excel (Exhibit 9-20).

$$v := \begin{bmatrix} 121 & 626 \\ 222 & 727 \\ 323 & 828 \\ 424 & 929 \\ 525 & 111 \end{bmatrix}$$

$$v = \begin{bmatrix} 121 & 626 \\ 222 & 727 \\ 323 & 828 \\ 424 & 929 \\ 525 & 111 \end{bmatrix}$$

<== Pasted to an Excel spreadsheet

121	626
222	727
323	828
424	929
525	111

Exhibit 9-24 Multiple columns of numbers copied to a Microsoft Excel worksheet (Exhibit 9-22).

Exhibit 9-25 Multiple column of numbers that will be copied from an output placeholder on a Mathcad worksheet (Exhibit 9-22.mcd).

Text from Microsoft Excel

Text can be transferred from Microsoft Excel by copying it into the clipboard and then pasting it onto a Mathcad worksheet. The text will be deposited as a text region onto the worksheet. See Exhibits 9-26 and 9-27 for the before and after text examples.

Text into Microsoft Excel

Text from a worksheet can be copied onto the clipboard and then pasted into a cell on a Microsoft Excel document. The text will appear as text on the Excel Spreadsheet. See Exhibits 9-28 and 9-29 for the before and after text examples.

Microsoft Access

Numbers from Microsoft Access

Records with numerical data from a Microsoft Access table or query can be copied onto the clipboard and then pasted onto a "Text Only" page from Notepad. The instructions below for data from Notepad should be followed.

Text to COPY into a Mathcad worksheet

Exhibit 9-26 This text was copied into a Mathcad worksheet (Exhibit 9-25).

Text to COPY into a Mathcad worksheet

Exhibit 9-27 This text was copied from a cell of a Microsoft Excel worksheet (Exhibit 9-24).

Text to copy into a Microsoft Excel Worksheet

Exhibit 9-28 This text was copied into a cell on a Microsoft Excel worksheet (Exhibit 9-27).

Text to copy into a Microsoft Excel Worksheet

Exhibit 9-29 This text was copied from a Mathcad worksheet (Exhibit 9-26).

Numbers into Microsoft Access

Numerical data can be transferred into Microsoft Access by copying the matrix into the clipboard and pasting it into an Access table. The number of columns of data must be at less than or equal to the number of columns in the Access table.

Microsoft Notepad

Numbers from Microsoft Notepad

Numerical data can be transferred from Microsoft Notepad into Mathcad simply by copying the text into the clipboard and then pasting it into an input placeholder on a Mathcad worksheet. Be careful; there can be no extraneous characters such as blanks, commas, or extra carriage returns included in the data being copied.

Multiple columns of data must be separated by tabs (or commas), with no extraneous characters such as blanks or extra carriage returns.

Numbers into Microsoft Notepad

Numerical data can be transferred into Microsoft Notepad by copying an array and then pasting it into your Notepad document. The pasted text will contain only the numbers and tabs as separators for the columns.

Text from Microsoft Notepad

Text can be transferred from Microsoft Notepad by copying it into the clipboard and then pasting it onto a Mathcad worksheet. The text will be deposited as a text region onto the worksheet.

Text into Microsoft Notepad

Text from a worksheet can be copied onto the clipboard and then pasted onto a Notepad document. The clipboard contains the rich text format of the text that is copied and uses the formatting when it pastes the text onto the document.

Microsoft WordPad

WordPad works the same as Word when it is working on a Word document. WordPad works the same as Notepad when it is working with an "unformatted document."

Microsoft Paint

Graphics from Microsoft Paint

Pictures from Microsoft Paint can be copied onto the clipboard and then pasted special onto a worksheet.

Graphics into Microsoft Paint

Pictures from a worksheet can be copied onto the clipboard and then pasted onto a Paint document.

Mathcad and the Internet

MathSoft Inc. (of Cambridge, Mass.), the developer of Mathcad, maintains a Web site (http://www.mathsoft.com) of general information about the company, its new products, technical support, product registration, product ordering, and e-mail to the company.

Mathcad helps you use Mathcad files from the Internet by providing a selection on the File menu ("Internet Setup") which helps you set up and access the Internet through Mathcad and your own Internet service provider. This allows you to go to Internet sites that contain Mathcad files and work with those files in their native environment.

The Mathcad 7 Web page (http://www.mathsoft.com/mathcad/) has helpful information and links to other sources and a mirror site containing much of the same information is also available for European users. The Mathcad site provides supplementary files that can be downloaded, mathematical puzzles, Mathcad graphics, on-line Mathcad journals, and links to sites with Mathcad solutions to practical problems.

The MathSoft Inc. Web page (http://www.mathsoft.com/index.html) has links to information about Mathcad and other math and graphing products.

Electronic Mail and Internet Mail

You exchange your Mathcad files (.MCD files) with friends and coworkers by electronic mail as follows.

If you use a Microsoft compatible e-mail system, you can send a Mathcad worksheet as an attachment to an e-mail message by opening the Mathcad worksheet and choosing "Send" from the File menu. Then just respond to the dialog boxes presented by your particular mail system.

For other e-mail systems you may have to encode or zip your Mathcad files and then attach them to your e-mail messages in order to mail them. The recipient will then have to unencode or unzip them before using them.

If you use America Online (AOL) for your e-mail system, you can send a Mathcad file to another AOL subscriber's address (or to certain other e-mail systems via the Internet) without any special preprocessing by simply attaching it to an e-mail message.

For other e-mail and Internet mail arrangements, you may have to experiment or talk to your e-mail administrator to determine what will work with your particular system.

Mathcad and Lotus Notes

You can use Lotus Notes to manage Mathcad worksheets by attaching them to a Lotus Notes database. The directory where Mathcad is installed has a file called "mathcad.ntf" which your Lotus Notes administrator can use in the creation of Mathcad databases. Your Notes administrator can acquaint you with the details of how Mathcad files can be used in your particular Notes installation.

Built-in Math Functions, and Programming

This final chapter of *Mastering Mathcad Version 7* presents a wide range of mathematical functions that can easily be inserted into your work and will discuss Mathcad's programming capabilities.

Inserting Math Functions into Your Work

In the preceding chapters we have used many of the powerful mathematical tools available in Mathcad. You have learned how to access these for your own work by selecting them from the palettes or typing them into your Mathcad documents. In addition to the many math functions we have seen previously in this book, there are still many more tools we have not yet encountered. The Insert menu contains a choice called "Function" that allows you select from over 200 math functions that Mathcad will then insert into your document at the cursor when you click on "Insert." All you need do is then enter the argument(s) in the placeholder(s). If you need help remembering exactly what that function does, or what types of arguments it takes, Mathcad also provides a brief description of the selected function.

If you enjoy browsing through reference books or math tables, you will probably be fascinated by the wealth of mathematical power these functions make available. You may see ones that will spark your curiosity and cause you to do some research or experimentation in order to be able to use them. You will see familiar trigonometric, exponential, and matrix functions such as you have already encountered. You will also find probability distributions and eigenvalues and some that you may not even recognize. You should remember that these functions can not only be used individually but can also be incorporated in your specially defined functions specific to your own work.

There are too many functions here for us to cover them all, and you will already be familiar with many of them. We list them below for reference purposes. Notice the number of arguments taken by the various functions. Some have a single argument, while others have multiple arguments. Names of Greek-letter functions are shown in English on our list. In Mathcad, these functions are in the order shown here, even though the Greek Letters are used. The available functions are:

acos(z)   The angle (in radians) whose cosine is z. Principal value for complex z.

acosh(z)   Inverse hyperbolic cosine. Principal value for complex z.

angle(x,y)   Angle (in radians) between the x-axis and the point (x, y). x and y must be real.

APPEND(file)   Appends a single value to "file.dat" on your disk.

APPENDPRN(file)   Appends a matrix to "file.prn" on your disk.

arg(z)   Angle (in radians) from the real axis to the complex number z.

asin(z)   The angle (in radians) whose sine is z. Principal value for complex z.

asinh(z)   Inverse hyperbolic sine. Principal value for complex z.

atan(z)   The angle (in radians) whose tangent is z. Principal value for complex z.

atanh(z)   Inverse hyperbolic tangent. Principal value for complex z.

augment(A,B)   A matrix formed by putting the two argument matrices side by side. A and B must have the same number of rows. Either or both arguments may be vectors.

bulstoer(v,x1,x2,acc,F,k,s)   A matrix of solution values for the differential equation specified by F having initial conditions v on the interval $[x1, x2]$ using a variable step Bulirsch-Stoer method. k and s govern the step-size.

Bulstoer(v,x1,x2,n,F)   A matrix of solution values for the differential equation specified by F having initial conditions v on the interval $[x1, x2]$ using a Bulirsch-Stoer method.

bvalfil(v1,v2,x1,x2,xi,F,L1,L2,S)   A set of initial conditions for the boundary value problem specified by F, $v1$, and $v2$ on the interval $[x1, x2]$ where the solution is known at the intermediate point xi.

ceil(x)   Smallest integer greater than or equal to x. x must be real.

CFFT(A)   Fast Fourier transform of complex data. Returns an array of the same size as its argument. Identical to cfft(A), except uses a different normalizing factor and sign convention.

cfft(A)   Fast Fourier transform of complex data. Returns an array of the same size as its argument.

cholesky(M)   The cholesky square root of the input matrix.

cnorm(x)   The integral from −infinity to x of the standard normal distribution. x must be real.

cols(A)   Number of columns in the array A.

concat(s1,s1)   Concatenation of two string arguments.

cond1(M)   The condition number of the input matrix based on the L1 norm.

cond2(M)   The condition number of the input matrix based on the L2 norm.

conde(M)   The condition number of the input matrix based on the Euclidean norm.

condi(M)   The condition number of the input matrix based on the infinity norm.

corr(A,B)   Correlation of the elements in A and B.

cos(z)   Cosine of z. z must be in radians.

cosh(z)   Hyperbolic cosine of z.

cot(z)   Cotangent of z. Argument in radians.

coth(z)   Hyperbolic cotangent of z.

csc(z)   Cosecant of z. Argument in radians.

csch(z)   Hyperbolic cosecant of z.

csort(A,n)   The input array, A, but with rows sorted so as to put column n in ascending order.

cspline(Mx,My)   Coefficients of cubic spline with cubic ends. Mx is an $n \times 1$ or $n \times 2$ matrix, and My is an $n \times 1$ or $n \times n$ matrix, respectively. The columns of Mx must be in ascending order. Returns a vector vs.

cvar(A,B)   Covariance of the elements in A and B.

dbeta(x,s1,s2)   The probability of selecting the value x of the beta distribution with shape parameters $s1$ and $s2$.

dbinom(x,sz,prob)   The probability of selecting a value x from the binomial distribution with size sz and probability of success prob.

dcauchy(x,l,s)   The probability of selecting a value x from the Cauchy distribution with location l and scale s.

dchisq(x,df)   The probability of selecting a value x from a Chi squared distribution with df degrees of freedom.

Delta(x,y)   Kronecker delta function. Returns 1 if $x = y$; otherwise, returns 0. x and y must be integers. To type Delta, press d+Ctrl+G.

dexp(x,r)   The probability of selecting a value x from the exponential distribution with rate of decay r.

dF(x,d1,d2)   The probability of selecting the value x from the F distribution with degrees of freedom $df1$ and $df2$.

dgamma(x,s)   The probability of selecting the value x of the gamma distribution with shape parameter s.

dgeom(x,prob)   The probability of selecting the value x from the geometric probability distribution with probability prob.

diag(v)   A diagonal matrix having the input vector as its main diagonal.

dlnorm(x,m,sd)   The probability that a value at x of the lognormal probability distribution with mean m and standard deviation sd.

dlogis(x,l,s)   The probability of selecting a value x from the logistic distribution with location l and scale s.

dnbinom(x,sz,prob)   The probability of selecting a value x from the negative binomial distribution with size sz and probability of failure prob.

dnorm(x,m,sd)   The probability of selecting the value x from the normal distribution with mean m and standard deviation sd.

dpois(x,lambda)   The probability of selecting a value x from the Poisson distribution with mean and standard deviation lambda.

dt(x,df)   The probability of selecting a value x from Student's t distribution with df degrees of freedom.

dunif(x,min,max)   The probability of selecting a value x from the uniform probability distribution with maximum max and minimum min.

dweibull(x,s)   The probability that a value at x of the Weibull distribution with shape parameter s.

eigenvals(M)   The eigenvalues of the input matrix.

eigenvec(M,z)   Eigenvector associated with eigenvalue z of the square matrix M.

eigenvecs(M)   A matrix whose columns constitute a set of eigenvectors for the input matrix.

Epsilon(i,j,k)   Completely antisymmetric tensor of rank three. i, j, and k must be integers between 0 and 2. Result is 0 if any two arguments are the same, 1 if the arguments are an even permutation of (0 1 2), and -1 if the arguments are an odd permutation of (0 1 2). To type Epsilon, press e+Ctrl+G.

erf(x)   Error function.

error(str)   Error message str.

exp(z)   Exponential: e raised to the power z.

FFT(v)   Fast Fourier transform of real data. v must be a real vector with 2^n elements, where n is an integer. Returns a vector of size 2^n-1 +1. Identical to fft(v), except uses a different normalizing factor and sign convention.

fft(v)   Fast Fourier transform of real data. v must be a real vector with 2^n elements, where n is an integer. Returns a vector of size $2^n - 1$ +1.

find(x1,x2,...)   Values of $x1$, $x2$, ... that solve a system of equations. Returns a scalar if only one argument, otherwise returns a vector of answers. This function must be preceded by guess values for each argument and the keyword "Given."

floor(x)   Greatest integer less than or equal to x. x must be real.

Gamma(z)   Euler's gamma function. To type Gamma, press G+Ctrl+G.

genfit(x,y,a0,F(x,a))   The parameters for the best fit by the nonlinear function defined by F to the x-y data given in the vectors x and y. The vector a0 gives initial guess for the parameters. F is a vector function of x and the parameters, with the fitting function in the first element and its derivative with respect to each parameter in the remaining elements.

geninv(A)   The generalized inverse of the input matrix.

genvals(M1,M2)   The eigenvalues for the generalized eigenvalue problem M1*v=lambda*M2*v

genvecs(M1,M2)   The eigenvalues for the generalized eigenvalue problem M1*v=lambda*M2*v

hist(intervals,data)   Histogram of data. intervals is an increasing real vector of interval limits. data is a real vector or matrix. Returns a vector of size one less than the size of intervals showing how many points of data fall in each interval.

I0(x)   Zeroth order modified Bessel function of the first kind. x must be real.

I1(x)   First order modified Bessel function of the first kind. x must be real.

ICFFT(A)   Inverse transform corresponding to CFFT. Returns an array of the same size as its argument.

icfft(A)   Inverse Fourier transform corresponding to cfft. Returns an array of the same size as its argument.

identity(n)   Identity matrix of size n. n must be a positive integer.

if(cond,x,y)   Either x or y depending on the value of cond. If cond is true (nonzero), the function returns x. If cond is false (zero), it returns y.

IFFT(v)   Inverse transform corresponding to FFT. Takes a vector of size $1+2^n-1$, where n is an integer. Returns a real vector of size 2^n.

ifft(v) Inverse transform corresponding to FFT. Takes a vector of size $1+2^n-1$, where n is an integer. Returns a real vector of size 2^n.

Im(z) Imaginary part of complex number z. Also works on vector and matrix arguments.

In(m,x) The mth order modified Bessel function of the first kind. x must be real. m must be between 0 and 100 inclusive.

intercept(vx,vy) Intercept of regression line. Takes two vector arguments vx and vy of same size. The elements of vx must be in ascending order. Returns a scalar: the y-intercept of the regression line.

interp(vs,Mx,My,x) Interpolated value from the coefficients vs. The vector vs is the output of one of the functions lspline, pspline, cspline, loess or regress. The matrices Mx and My are the data inputs to that function and x is a $1 \times n$ vector specifying the value at which to evaluate the interpolating function.

iwave(v) The inverse 1-dimensional discrete wavelet transform of v computed using the Daubechies 4-coefficient wavelet filter.

J0(x) Zeroth order Bessel function of the first kind. x must be real.

J1(x) First order Bessel function of the first kind. x must be real.

Jn(m,x) The mth order Bessel function of the first kind. x must be real. m must be between 0 and 100 inclusive.

K0(x) Zeroth order modified Bessel function of the second kind. x must be real and positive.

K1(x) First order modified Bessel function of the second kind. x must be real and positive.

Kn(m,x) The mth order modified Bessel function of the second kind. x must be real and positive. m must be between 0 and 100 inclusive.

kronecker(M,N) Kronecker product of a square matrix M and a square matrix N.

ksmooth(vx,vy,bw) Gaussian kernel smoothing with bandwidth bw of the scatterplot of vy versus vx.

last(v) Index of last element in vector v. Returns a scalar.

length(v) Number of elements in vector v. Returns a scalar.

linfit(x,y,F) The coefficients for the linear combination of the functions in F that best fit the x-y data given in the vectors x and y.

linterp(vx,vy,span) Linearly interpolated value. Takes two vector arguments vx and vy of same size and a scalar x at which to interpolate, returns a scalar. The elements of vx should be in ascending order.

ln(z) Natural logarithm (base e) of z. Returns principal value (imaginary part between pi and $-$pi for complex z).

loess(Mx,vy,x) Local regression of vy versus Mx with neigborhood size parameter span.

log(z) Common logarithm (base 10) of z.

lsolve(M,v) The solution to the linear system M*x=v.

lspline(Mx,My) Coefficients of cubic spline with linear ends. Mx is an $n \times 1$ or $n \times 2$ matrix and My is an $n \times 1$ or $n \times n$ matrix respectively. The columns of Mx must be in ascending order. Returns a vector vs.

lu(M) The matrices P, L, and U forming the LU decomposition of the input matrix P*M=L*U.

matrix(m,n,f)   An mxn matrix in which each element is given by $f(i,j)$ where: $i=0.. \, m-1$ and $j=0.. \, n-1$.

max(A)   Largest element in A. If A is complex, returns $\max(\text{Re}(A)) + i*\max(\text{Im}(A))$.

mean(A)   Mean of elements of A. Returns a scalar.

median(A)   Median of the elements in A.

medsmooth(v,n)   The running median-of-n smoothing of the data vector v.

min(A)   Smallest element in A. If is A complex, returns $\min(\text{Re}(A)) + i*\min(\text{Im}(A))$.

minerr(x1,x2,...)   Values of $x1$, $x2$, ... that come closest to solving a system of equations. Returns a scalar answer if only one argument is given, otherwise returns a vector of answers. This function must be preceded by guess values for each argument as well as the keyword "Given".

mod(x,modulus)   Remainder on dividing x by modulus. Arguments must be real. Result has same sign as x.

muntigrid(M,n)   A matrix of solution values for Poisson's equation where the solution is assumed to be 0 on the boundary.

norm1(M)   The L1 norm of the input matrix.

norm2(M)   The L2 norm of the input matrix.

norme(M)   The Euclidean norm of the input matrix.

normi(M)   The infinity norm of the input matrix.

num2str(n)   Converts a number to a string.

pbeta(q,s1,s2)   The probability that a value from a beta distribution with shape parameters $s1$ and $s2$ will be less than q.

pbinom(q,sz,prob)   The probability that a value from a binomial distribution with size sz and probability of success prob will be less than q.

pcauchy(q,l,s)   The probability that a value from a Cauchy distribution with location l and scale s will be less than q.

pchisq(q,df)   The probability that a value from a Chi squared distribution with df degrees of freedom will be less than q.

exp(q,r)   The probability that a value from an exponential distribution with rate of decay r will be less than q.

pF(q,df1,df2)   The probability that a value from an F distribution with degrees of freedom df1 and df2 will be less than q.

pgamma(q,s)   The probability that a value from an F distribution with degrees of freedom df1 and df2 will be less than q.

pgeom(q,prob)   The probability that a value from a geometric distribution with probability prob will be less than q.

Phi(x)   Heaviside step function: 1 if x is greater than or equal to 0 and 0 otherwise. To type Phi, press F+Ctrl+G.

plnorm(q,m,sd)   The probability that a value from a lognormal distribution with mean m and standard deviation sd will be less than q.

plogis(q,l,s)   The probability that a value from a logistic distribution with location l and scale s will be less than q.

pnbinom(q,sz,prob)   The probability that a value from a negative binomial distribution with size sz and probability of failure prob will be less than q.

pnorm(q,m,sd)   The probability that a value from a normal distribution with mean m and standard deviation sd will be less than q.

polyroots(C)   A vector containing all the complex roots of the polynomial with complex coefficients C.

ppois(q,lambda)   The probability that a value from a Poisson distribution with mean and standard deviation lambda will be less than q.

predict(v,m,nfut)   A vector of nfut values predicted from the data v using m linear prediction coefficients.

Psi(z)   Digamma function for complex z.

pspline(Mx,My)   Coefficients of cubic spline with parabolic ends. Mx is an $n \times 1$ or $n \times 2$ matrix, and My is an $n \times 1$ or $n \times n$ matrix, respectively. The columns of Mx must be in ascending order. Returns a vector vs.

pt(q,df)   The probability that a value from Student's t distribution with df degrees of freedom will be less than q.

punif(q,min,max)   The probability that a value from a uniform distribution with minimum min and maximum max will be less than q.

pweibull(q,s)   The probability that a value from a Weibull distribution with shape parameter s will be less than q.

qbeta(p,s1,s2)   The value q such that there is a probability p that a sample from a beta distribution with shape parameters $s1$ and $s2$ will be less than q.

qbinom(p,sz,prob)   The value q such that there is a probability p that a sample from a binomial distribution with size sz and probability of success prob will be less than q.

qcauchy(p,l,s)   The value q such that there is a probability p that a sample from a Cauchy distribution with location l and scale s will be less than q.

qchisq(p,df)   The value q such that there is a probability p that a sample from a Chi squared distribution with df degrees of freedom will be less than q.

qexp(p,sd)   The value q such that there is a probability p that a sample from an exponential distribution with rate of decay r will be less than q.

qF(p,df1,df2)   The value q such that there is a probability p that a sample from an F distribution with degrees of freedom $df1$ and $df2$ will be less than q.

qgamma(p,s)   The value q such that there is a probability p that a sample from a gamma distribution with shape parameter s will be less than q.

qgeom(p,prob)   The value q such that there is a probability p that a sample from a geometric distribution with probability prob will be less than q.

qlnorm(p,m,ssd)   The value q such that there is a probability p that a sample from a lognormal distribution with mean m and standard deviation sd will be less than q.

qlogis(p,l,s)   The value q such that there is a probability p that a sample from a logistic distribution with location l and scale s will be less than q.

qnbinom(p,sz,prob)   The value q such that there is a probability p that a sample from a negative binomial distribution with size sz and probability of failure prob will be less than q.

qnorm(p,m,sd)   The value q such that there is a probability p that a sample from a normal distribution with mean m and standard deviation sd will be less than q.

qpois(p,lambda)   The value q such that there is a probability p that a sample from a Poisson distribution with mean and standard deviation lambda will be less than q.

qr(A)   The matrices Q and R forming the QR decomposition of the input matrix Q*R=A.

qt(p,df)   The value q such that there is a probability p that a sample from Student's t distribution with df degrees of freedom will be less than q.

qunif(p,min,max)   The value q such that there is a probability p that a sample from a uniform distribution with maximum max and minimum min will be less than q.

qweibull(p,s)   The value q such that there is a probability p that a sample from a Weibull distribution with shape parameter s will be less than q.

rank(A)   The rank of the input matrix.

rbeta(n,s1,s2) n independent random numbers from a beta distribution with shape parameters $s1$ and $s2$.

rbinom(n,sz,prob) n independent random numbers from a binomial distribution with size sz and probability of success prob.

rcauchy(n,l,s) n independent random numbers from a Cauchy distribution with lcoation l and scale s.

rchisq(n,df) n independent random numbers from a Chi squared distribution with df degrees of freedom.

Re(z)   Real part of complex number z.

READ(file)   A single value taken from "file.dat" on your disk.

READ_BLUE(F)   A matrix representing the RGB blue values of the color image file F.

READ_GREEN(F)   A matrix representing the RGB green values of the color image file F.

READ_HLS(F)   A packed matrix of HLS values for the image file F. The returned matrix is actually 3 matrices (H,L,S) packed side by side and represents the Hue, Saturation, and Lightness values based on the Ostwald color model.

READ_HLS_HUE(F)   A matrix representing the HLS hues of the image file F. The returned matrix contains the Hue values based on the Ostwald color model.

READ_HLS_LIGHT(F)   A matrix representing the HLS Lightness of the image file F. The returned matrix contains the Lightness values based on the Ostwald color model.

READ_HLS_SAT(F)   A matrix representing the HLS Saturation of the image file F. The returned matrix contains the Saturation values based on the Ostwald color model.

READ_HSV (F)   A packed matrix of HSV values for the image file F. The returned matrix is actually three matrices (H,S,V) packed side by side and represents the Hue, Saturation, and Value values based on the Smith's HSV color model.

READ_HSV_HUE(F)   A matrix representing the HSV hues of the image file F. The returned matrix contains the Hue values based on Smith's HSV color model.

READ_HSV_SAT(F)   A matrix representing the HSV Saturation of the image file *F*. The returned matrix contains the Saturation values based on Smith's HSV color model.

READ_HSV_VALUE(F)   A matrix representing the HSV Value of the image file *F*. The returned matrix contains the Value values based on Smith's HSV color model.

READ_IMAGE(im)   Reads the BMP file im as a greyscale image.

READ_RED(F)   A matrix representing the RGB red values of the color image file *F*.

READBMP(im)   Reads the BMP file im as a greyscale image.

READPRN(file)   A matrix taken from the structured data file "file.prn" on your disk.

READRGB(im)   Reads the BMP file im as an RGB color image into a matrix.

regress(Mx,vy,n)   The coefficients for the multivariate degree *n* least squares polynomial fit of the response data *vy* versus the factors *Mx*.

relax(M1,M2,M3,M4,M5,A,U,x)   A matrix of solution values for Poisson's equation where *A* contains values for the source term and *U* specifies the initial boundary condition and approximate interior values.

reverse(A)   The input array, *A*, but with the order of the elements in each column reversed.

rexp(n,r)   *n* independent random numbers from an exponential distribution with rate of decay *r*.

rF(n,df1,df2)   *n* independent random numbers from an *F* distribution with degrees of freedom *df1* and *df2*.

rgamma(n,s)   *n* independent random numbers from a gamma distribution with shape parameter *s*.

rgeom(n,prob)   *n* independent random numbers from a geometric distribution with probability prob.

rkadapt(v,x1,x2,acc,F,k,s)   A matrix of solution values for the differential equation specified by *F* having initial conditions *v* on the interval [*x1*, *x2*] using a variable step Runge-Kutta method. *k* and *s* govern the step-size.

Rkadapt(v,x1,x2,n,F)   A matrix of solution values for the differential equation specified by *F* having initial conditions *v* on the interval [*x1*, *x2*] using an adaptive step Runge-Kutta method.

rkfixed(v,x1,x2,n,F)   A matrix of solution values for the differential equation specified by *F* having initial conditions *v* on the interval [*x1*, *x2*] using a fixed step Runge-Kutta method.

rlnorm(n,m,sd)   *n* independent random numbers from a lognormal distribution with mean *m* and standard deviation *sd*.

rlogis(n,l,s)   *n* independent random numbers from a logistic distribution with location *l* and scale *s*.

rnbinom(n,sz,prob)   *n* independent random numbers from a negative binomial distribution with size *sz* and probability of failure prob.

rnd(x)   Uniformly distributed random number between 0 and *x*. *x* is real.

rnorm(n,m,sd)   *n* independent random numbers from a normal distribution with mean *m* and standard deviation *sd*.

root(expr,var)   Value of var at which expr is zero. This function must be preceded by a guess value for var.

rows(A)   Number of rows in the array *A*.

rpois(n,lambda)   *n* independent random numbers from a Poisson distribution with mean and standard deviation lambda.

rref(A)   The row-reduced echelon form of the input matrix.

rsort(A,n)   The input array, A, but with columns sorted so as to put row n in ascending order.

rt(n,df) n independent random numbers from Student's t distribution with df degrees of freedom.

runif(n,min,max) n independent random numbers from a uniform distribution with minimum min and maximum max.

rweibull(n,s) n independent random numbers from a Weibull distribution with shape parameter s.

sbval(v,x1,x2,F,L,S)   A set of initial conditions for the boundary value problem specified by F and v on the interval $[x1, x2]$.

search(s1,s2,n)   Searches $s1$ for the first occurrence of $s2$ starting at character i. Returns the character position if found, -1 otherwise.

sec(z)   Secant of z. Argument in radians.

sech (z)   Hyperbolic secant of z.

sin(z)   Sine of z. Argument in radians.

sinh(z)   Hyperbolic sine of z.

slope(vx,vy)   Slope of regression line. vx and vy are vectors having the same number of elements. The elements of vx must be in ascending order.

sort(v)   The input vector, v, but with its elements sorted in ascending order.

stack(A,B)   Matrix A stacked on top of matrix B.

Stdev(A)   The square root of the unbiased sample variance of the data vector v.

stdev(A)   Standard deviation of the elements in A.

stiffb(v,x1,x2,acc,F,J,k,s)   A matrix of solution values for the stiff differential equation specified by F and the Jacobian function J, having initial conditions v on the interval $[x1, x2]$ using a variable step Bulirsch-Stoer method. k and s govern the step-size.

Stiffb(v,x1,x2,n,F,J)   A matrix of solution values for the stiff differential equation specified by F and the Jacobian function J, having initial conditions v on the interval $[x1, x2]$ using a Bulirsch-Stoer method.

stiffr(v,x1,x2,acc,F,J,k,s)   A matrix of solution values for the stiff differential equation specified by F and the Jacobian function J, having initial conditions v on the interval $[x1, x2]$ using a variable step Rosenbrock method. k and s govern the step-size.

Stiffr(v,x1,x2,n,F,)   A matrix of solution values for the stiff differential equation specified by F and the Jacobian function J, having initial conditions v on the interval $[x1, x2]$ using a Rosenbrock method.

str2num(s)   Converts a string to a number.

str2vec(s)   Converts a string to a vector of ascii codes.

strlen(s)   String length.

submatrix(M,nrl,nrh,ncl,nch)   Extracts a submatrix from M.

substr(s,m,n)   Substring of s starting at character m and of length n.

supsmooth(vx,vy)   Variable bandwidth smoothing of the scatterplot of vy versus vx.

svd(A)   A matrix containing the complete singular value decomposition of the input matrix.

svds(A)   The vector of singular values of the input matrix.

tan(z)   Tangent of z. Argument in radians.

tanh(z)   Hyperbolic tangent of z.

tr(M)   Trace of the square matrix M: sum of diagonal elements.

until(x,y)   Returns y until x is negative.

Var(A)   The unbiased sample variance of the data vector v.

var(A)   Variance of the elements in A.

vec2str(v)   Converts a vector of ascii codes to a string.

wave(v)   The 1-dimensional discrete wavelet transform of v computed using the Daubechies 4-coefficient wavelet filter.

WRITE(file)   Writes a single value to "file.dat" on your disk.

WRITE_HLS(file)   An HLS packed matrix written out to a Windows 16 million color bitmap file. The file is named "file.bmp".

WRITE_HSV(file)   An HSV packed matrix written out to a Windows 16 million color bitmap file. The file is named "file.bmp".

WRITEBMP(M)   Writes the matrix M as a greyscale image in a Windows BMP format.

WRITEPRN(file)   Writes an array into "file.prn" on your disk.

WRITEGE(M)   Writes the mx3n matrix M as an RGB color image in Windows BMP format.

Y0(x)   Zeroth order Bessel function of the second kind. x must be real and positive.

Y1(x)   First order Bessel function of the second kind. x must be real and positive.

Yn(m,x)   The mth order Bessel function of the second kind. x must be real and positive. m must be between 0 and 100 inclusive.

While we cannot cover all of these, we will select a few examples to give an idea of how you might use them. These functions are described in detail in the "Built-in functions listed alphabetically" section in the Appendix of the *Mathcad User's Guide*.

In Exhibit 10-1 we see examples of functions that have been selected from the Insert and Function menus and for which the appropriate argument (or arguments) have been entered. A brief explanation of each function is also given. Many other functions can be entered into Mathcad documents with equal ease.

In Exhibit 10-2 we see examples of how to use Mathcad's functions to create your own, user-defined functions. All you need to do is insert them into your expressions and enter values into the placeholders. We encourage you to try using the ones you already know about and to learn about the others. Mathcad facilitates experimentation and learning, and there is no penalty for getting something wrong when you experiment.

Programming

Mathcad mostly makes it possible to do your work without writing programs. But for the times you need to do programming, Mathcad makes a powerful programming capability available.

$\Gamma(5) = 24$ Returns Euler's gamma function of argument. $\Gamma(n)=(n-1)!$

                                                    Another example

                                                    $\Gamma(23) = 1.1240007278 \cdot 10^{21}$
                                                    $22! = 1.1240007278 \cdot 10^{21}$

$Y0(5) = -0.309$ For positive argument x, returns Bessel function
                                                    $Y_0(x)$.

$\delta(4,6) = 0$ Kronecker delta function, returns 1 if the two arguments are
                                                    equal, 0 otherwise.

                                                    Another example

                                                    $\delta(7,7) = 1$

$rnd(10) = 0.013$ For a real argument x returns a random number
                                                    between 0 and x.

                                                    $rnd(10) = 1.933$

                                                    $rnd(10) = 5.85$

$mod(69,9) = 6$ Remainder on dividing first argument by the second.
                                                    Arguments must be real. Answer has same sign as first
                                                    argument.

Exhibit 10-1 Examples of functions.

In a way, this section should really be called "advanced programming" because everything that has been performed so far has actually been programming in some sense.

"When the only tool you own is a hammer, every problem begins to resemble a nail."
ABRAHAM MASLOW

And so it is with any programming language! Programmers usually think in terms of the language constructs available to them in order to solve complex problems. Most programmers have a favorite language such as COBOL, FORTRAN Visual Basic, or C++. These languages are fine if you need to solve a wide range of general problems; however, everything necessary

Example 1

$a := 5$

$f(x) := x^4 \cdot \ln(x)$

$f(a) = 1.006 \cdot 10^3$

Example 2

$g(x) := \ln(x) + x$

$g(6) = 7.792$

Example 3

$h := rnd(10) - rnd(10)$

$h = 6.487$

$h := rnd(10) - rnd(10)$

$h = 4.065$

$h := rnd(10) - rnd(10)$

$h = -0.559$

Exhibit 10-2

to solve almost any mathematical problem is provided by Mathcad. The problem with a favorite language is that you tend to think of solutions in terms of that language—in other words, programmers write programs. With Mathcad you want to think of solutions in terms of mathematics. You normally don't write programs in Mathcad.

It's one thing to be able to solve a problem, and another to validate the solution to be sure that it will give correct answers under all conditions under which it will apply. Users are responsible for the accuracy of their own program's answers.

The tools used for programming in Mathcad are provided by displaying the Programming palette. The Programming palette is enabled by clicking on the Programming icon which is

Exhibit 10-3 Palette of palettes.

Add Line	←	if
while	for	break
otherwise	return	on error
continue		

Exhibit 10-4 Programming button and palette.

| Add Line |

Exhibit 10-5 Add Line button.

found in the palette of Palettes shown in Exhibit 10-3. The Programming button and palette are shown in Exhibit 10-4.

You start a program by defining a function such as "f(x) :=" and then clicking the "Add Line" button shown in Exhibit 10-5. This will create the vertical line that defines a program as shown in Exhibit 10-6. This normally is performed while the left-hand side of an assignment operator of a function is selected. Programs start out as two lines initially. Programs can be as many lines long as you desire, by just adding lines with the Add Line button. Do not type the words "Add Line."

The function of x is given the value of the very last executed assignment statement; in the case shown in Exhibit 10-6, the value will be set to 2.

There are four elementary programming constructs: "assignment," "if," "for," and "while." These constructs provide the logic for all the advanced programming. These will guarantee "structured" programming because there is no "go to" statement as in most other languages.

$$f(x) := \begin{vmatrix} \blacksquare \\ \blacksquare \end{vmatrix}$$

$$f(x) := \begin{vmatrix} 0 \\ x \\ 2 \end{vmatrix}$$

Exhibit 10-6 Add Line operator.

$$f(x) := \begin{vmatrix} a \leftarrow 5 \\ x \\ b \leftarrow 6 \end{vmatrix}$$

Exhibit 10-7 Local Assignment button.

Exhibit 10-8

Exhibit 10-9 "if" programming statement button.

You use only those statements needed to perform the proper type of operations required to solve your problem. These four powerful statements are all you should ever need.

The first statement needed in a program is an assignment statement. The normal "Mathcad :=" assignment statement cannot be used within a program. The "local assignment" (left arrow) button from the programming palette is used for assignments within a program (see Exhibit 10-7). This helps emphasize that the variables being assigned are local variables within the program. Any variable created within a program is local only within that program and not available outside the program. Results of a program are, of course, available externally as the results.

In Exhibit 10-8, the function of x is given the value of the last executed assignment statement, in this case 6. The variables "a" and "b" are local only within the program and remain undefined if they were undefined before the program or would retain the value which they had before the program was executed.

The second statement used for programming is the "if" statement, which is activated with the "if" button on the programming palette while a programming line is selected (see Exhibit 10-9). Do not type the word "if" unless you are using an "if" as a variable within a statement (see Exhibit 10-10). If x is given a value of 5 when the function is called, then the value returned is 6; otherwise the last statement for the assignment of 6 is not executed.

Another useful statement is the "otherwise" statement (see "otherwise" button in Exhibit

$$f(x) := \begin{vmatrix} a \leftarrow 5 \\ x \\ 6 \ \text{if} \ \blacksquare \end{vmatrix}$$

$$f(x) := \begin{vmatrix} a \leftarrow 5 \\ x \\ 6 \ \text{if} \ a = x \end{vmatrix}$$

Exhibit 10-10

$$Crn(N) := \begin{vmatrix} 1 & \text{if } N \leq 3 \\ Crn(N-1) + Crn(N-2) + Crn(N-3) & \text{otherwise} \end{vmatrix}$$

$$k := 1 \mathinner{.\,.} 12$$

k	Crn(k)
1	1
2	1
3	1
4	3
5	5
6	9
7	17
8	31
9	57
10	105
11	193
12	355

otherwise

Exhibit 10-11 "otherwise" statement button.

Exhibit 10-12

10-11); "otherwise" is used only after an "if" statement (see Exhibit 10-12). Use the "otherwise" button on the Programming palette; do not type the word "otherwise." In Exhibit 10-12 either the last line or the next-to-last line is executed. The function is given the value of the last statement executed, not the value of the last statement in the program, because only the "if" or the "otherwise" will be executed.

The last two statements are both loop instructions. The first of these two is the "for" statement, which stops the loop after a fixed number of loops; the other is the more general "while" statement, which uses a logical expression to determine when to stop. Both loops can stop early using a "break" statement.

The "for" statement executes a loop for a fixed number of times specified by a range of values. The "for" statement is created by clicking the "for" button on the programming palette (see Exhibit 10-13). Do not type the word "for." The "for" statement generates the three

for

Exhibit 10-13 "for" loop statement button.

for $i \in 1, 2 .. 7 = 5.444$

$$\begin{vmatrix} a \leftarrow \dfrac{i}{3} \\ b \leftarrow a^2 \end{vmatrix}$$

for $\blacksquare \in \blacksquare$

$\blacksquare$

Exhibit 10-14 "for" loop operator.

Exhibit 10-15 This will set "i" to the values 1 through 7 by 1 and evaluate the loop for each value.

---

placeholders as in Exhibit 10-14. The first placeholder (on the left of symbol) is for the iteration variable. The second placeholder to the right is for the range to be given to the iteration variable. The third placeholder is for the program that is to be executed within the loop. More lines can be added to the loop by using the Add Line button.

Exhibit 10-15 shows an example of a "for" loop. Any variables created within a loop are local, the same as with all programs. The loop is given the last value executed in the loop.

The "while" statement executes a loop until a given condition is no longer TRUE. The "while" statement can exist only within a program. The "while" statement is created by clicking the "while" button on the Programming palette (see Exhibit 10-16). Do not type the word "while." The "while" statement generates the two placeholders as in Exhibit 10-17. The first placeholder (on the "while" line) is for the conditional statement. The second placeholder is for the program that is to be executed while the condition is true. More lines can be added to the loop by using the Add Line button. The condition must originally be TRUE, or else the "while" will not execute at all. Be aware that the condition that was originally TRUE must be changed within the loop or the "while" will loop continuously (or until you hit the "escape" key).

Exhibit 10-18 shows an example of a "while" loop. The loop is given the last value executed in the loop. In this example "A" is set to 5, and the counter "I" is set to 0. The "while" condition is "I < A," which is definitely true at the start. The "while" loop consists of incrementing "I" by 1 and then calculating 2 raised to the "I" power. The "while" stops when "I" is 5, because we increment before we test the condition and the result is 32 "(2^5)."

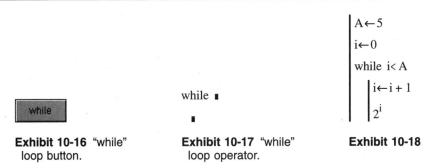

while $\blacksquare$

$\blacksquare$

$$\begin{vmatrix} A \leftarrow 5 \\ i \leftarrow 0 \\ \text{while } i < A \\ \begin{vmatrix} i \leftarrow i + 1 \\ 2^i \end{vmatrix} \end{vmatrix}$$

Exhibit 10-16 "while" loop button.

Exhibit 10-17 "while" loop operator.

Exhibit 10-18

Exhibit 10-19 "break"
statement button.

The "break" statement is used to stop a "for" loop, a "while" loop, a subprogram, or a program. The "break" statement is created by clicking the "break" button on the programming palette (see Exhibit 10-19). Do not type the word "break." The "break" statement generates no place-holders as shown in Exhibit 10-20.

A "break" statement in a "while" or "for" loop causes the flow to stop only within that loop. The loop has already been assigned a value when the last statement prior to the break was executed. This value is passed up to the level of the program which executed the loop.

A "break" statement in a program (other than a "while" or a "for" loop) causes the flow to stop within the level that is currently executing and each calling program level above it until it reaches either the highest level, a "while" loop, or a "for" loop. The level with the "break" has already been assigned a value when the last statement prior to the break was executed. This value is passed up the line to either the highest level which stops, a "while" loop which stops, or a "for" loop which stops.

Now you have the programming tools to tackle virtually any programming you might need to do in Mathcad. However, you will see in the next example, which is adapted from a real programming problem, that Mathcad has a built-in mathematical solver, which often renders programming unnecessary. For example, the following problem was originally solved by a long and complex FORTRAN program with the usual loops, branches, and subroutines.

You could do a straightforward translation of this problem from FORTRAN into Mathcad programming, using the programming tools we have already described. Such a translation would be approximately 20 pages long and would be basically similar to a FORTRAN program, except that the statements would be Mathcad, not FORTRAN.

Instead, we show here a concise solution (the original was a single page; we have lengthened it here to produce a clearer explanation of the method). The key to such a concise, elegant solution is Mathcad's powerful "solve" capability, which for many problems allows you to set up the mathematics and turn the solving over to Mathcad—thus avoiding programming.

Pulley Programming Example

This pulley, or belt adjustment, program example was originally written in FORTRAN some years ago by one of the authors. It was written to replace the manual manipulations and

---

break

Exhibit 10-20 "break"
operator.

calculations a draftsperson needed to perform to manually solve this problem on a drawing board. Prior to the program, draftspersons iterated on drawings, and measured the length of the belt, using a map measure. The FORTRAN program was long and cumbersome with many logic statements that guarded against dividing by zero. It contained a routine that used a Secant iteration technique to compute the belt length. The advantage of using Mathcad is that it normally does not require programming. You simply solve the mathematics. It can also give instant visual feedback by using graphics.

The complete solution to the slot adjustment angle for three pulleys is given in the Appendix to this book (Exhibit A-1). It would not be very difficult to add a fourth pulley to the system if that were appropriate. Several important programming practices are demonstrated in the example:

1. Highlight the input variables that are kept at the beginning of the problem and grouped by type.

2. Display your problem graphically, if practical, for visual verification. (See Exhibit A-2 in the Appendix.)

3. Validate your results to be sure the answer is reasonable and within proper bounds.

4. Don't think in terms of a programming language such as FORTRAN. The first attempt to translate this problem from FORTRAN required 12 pages, three subroutines, and a "while" loop.

Exhibit 10-21 is the first page of our programming example. It gives a brief description of the problem. An engineer determines the ideal location for the accessory drive on the engine. That is the starting point for the determination of the minimum belt length. The minimum belt length is rounded up to be a multiple of the belt length tolerance. One tolerance is added to the belt length, and this becomes the nominal belt length. That is the length of the belt that the owner would buy to replace the belt. The slot for the pulley adjustment must accommodate not only a belt one tolerance smaller than nominal but also a belt that is one tolerance longer that has stretched to its limit.

Exhibit 10-22 shows the input variables. The input variables have all been highlighted on the CD. This is the ideal starting position for the accessory pulley and the pivot arm. A maximum allowable slot angle is used to keep the size of the arrays and the processing time

---

When designing accessory mounting brackets for automotive engines weight is a major consideration. This problem is to determine the minimum adjustment angle for an accessory pulley with a belt which has a tolerance of +/- 0.25 inches and a 2% stretch limit.

Pulley 3 is initially positioned as close to the engine as practical and the pivot point is chosen so as to allow the maximum belt stretch to be "taken up" with the adjustment slot for pulley 3. The smaller the slot angle, the less the weight. Once the pivot point and the initial pivot angle are defined, the belt length and slot angle must be found.

Exhibit 10-21 Belt adjustment program.

Belt Dimensions

Tolerance := .25   BeltStretch := .02   Percent of Length

Engine Dimensions

r1 := 4   Pulley radius   X := 7   Pivot Location

r2 := 3   Pulley radius   Y := 6   Pivot Location

r3 := 2   Pulley radius   L := 3   Pivot Arm Length

V := 10   Vertical distance Pulley 1 to 2   OriginalAngle := 75   Pivot Arm Degrees

MaxSlotAngle := -10.   Maximum allowable angular position - in degrees

Exhibit 10-22

---

to a minimum. Notice that the MaxSlotAngle is −10 because zero degrees (0°) is horizontal and this specifies 10° below horizontal.

Exhibit 10-23 shows the three pulleys at the starting position on the engine. Pulley 3 starts as close to the engine as practical, and the direction of the adjustment is portrayed with a dashed arc rotating clockwise. This picture is from Exhibit A-2 in the Appendix, which is the

---

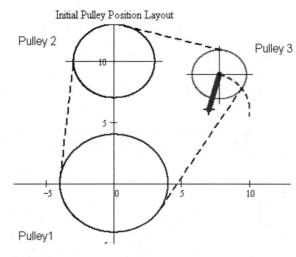

Exhibit 10-23 Initial pulley layout showing the pivot arm maximum allowable swing motion.

Angle := OriginalAngle·deg   radians

Angle = 1.309   radians

Exhibit 10-24

$$\text{Length}(A) := \text{Wrap1}(A)\cdot r1 + \text{Len12} + \text{Wrap2}(A)\cdot r2 + \text{Len23}(A) + \text{Wrap3}(A)\cdot r3 + \text{Len31}(A)$$

Exhibit 10-25

---

pulley example with drawings. If there is a problem with the picture, then either the input is erroneous or the program may have a problem. Hopefully it is the former, but the good news is that you will be visually alerted to the problem.

In Exhibit 10-24 the starting angle for the pivot arm (also the slot angle) is converted to radians and displayed. The next step is to determine the length of a belt that will fit around all three pulleys. Exhibit 10-25 shows the function that will be used to compute the length. The parameter "A" is the current slot (pivot arm) angle. It is the sum of the length of belt that is wrapped around each individual pulley (wrap angle times the radius) and the length of belt that is needed between each pair of pulleys (tangent to each pulley).

The functions for the computation of the wrap angles are shown in Exhibit 10-26. The parameter "A" is the current slot angle. The functions for the computation of the belt length are shown in Exhibit 10-27. The length of belt between pulleys 1 and 2 is constant. Exhibit 10-28 shows the calculation of the minimum belt length and its value. Exhibit 10-29 shows

---

$$\text{Wrap1}(A) := \frac{\pi}{2} + \text{atan}\left(\frac{r1 - r2}{V}\right) - \text{atan}\left[\frac{r3 - r1}{\sqrt{(X + L\cdot\cos(A))^2 + (Y + L\cdot\sin(A))^2}}\right] + \text{atan}\left(\frac{Y + L\cdot\sin(A)}{X + L\cdot\cos(A)}\right)$$

$$\text{Wrap2}(A) := \frac{\pi}{2} + \text{atan}\left[\frac{r2 - r3}{\sqrt{(X + L\cdot\cos(A))^2 + (V - Y - L\cdot\sin(A))^2}}\right] - \text{atan}\left(\frac{r1 - r2}{V}\right) \ldots$$
$$+ \text{atan}\left(\frac{V - Y - L\cdot\sin(A)}{X + L\cdot\cos(A)}\right)$$

$$\text{Wrap3}(A) := \left[\left[\left[\pi + \text{atan}\left[\frac{r3 - r1}{\sqrt{(X + L\cdot\cos(A))^2 + (Y + L\cdot\sin(A))^2}}\right] \ldots\right] \ldots\right] - \text{atan}\left(\frac{Y + L\cdot\sin(A)}{X + L\cdot\cos(A)}\right)\right.$$
$$+ -\text{atan}\left[\frac{r2 - r3}{\sqrt{(X + L\cdot\cos(A))^2 + (V - Y - L\cdot\sin(A))^2}}\right]$$
$$+ -\text{atan}\left(\frac{V - Y - L\cdot\sin(A)}{X + L\cdot\cos(A)}\right)$$

Exhibit 10-26

$$\text{Len12} := \sqrt{V^2 - (r1 - r2)^2}$$

$$\text{Len23(A)} := \sqrt{X^2 + 2 \cdot X \cdot L \cdot \cos(A) + (V - Y)^2 - 2 \cdot V \cdot L \cdot \sin(A) + 2 \cdot Y \cdot L \cdot \sin(A) + L^2 - (r2 - r3)^2}$$

$$\text{Len31(A)} := \sqrt{X^2 + 2 \cdot X \cdot L \cdot \cos(A) + Y^2 + 2 \cdot Y \cdot L \cdot \sin(A) + L^2 - (r3 - r1)^2}$$

Exhibit 10-27

$$\text{MinimumBeltLength} := \text{Length(Angle)} \quad \text{Get minimum belt length}$$

$$\text{MinimumBeltLength} = 49.073$$

Exhibit 10-28

$$\text{NominalBeltLength} := \text{ceil}\left(\frac{\text{MinimumBeltLength}}{\text{Tolerance}} + 1\right) \cdot \text{Tolerance} \qquad \text{Round length up to tolerance}$$

$$\text{NominalBeltLength} = 49.5$$

Exhibit 10-29

$$\text{MaxBeltLength} := (\text{NominalBeltLength} + \text{Tolerance}) \cdot (1.0 + \text{BeltStretch}) \qquad \text{Apply tolerance and stretch}$$

$$\text{MaxBeltLength} = 50.745$$

Exhibit 10-30

Angle := (OriginalAngle − 3.)·deg   Guess Value

Given

Length(Angle) = MaxBeltLength   Equation being solved

SlotAngle := Find(Angle)   <==== If this evaluates to "did not find solution" then change the geometry so that the MaxBeltLength is less than MaxLen shown below. Unfortunately pulley 3 can move beyond the position that yields the greatest belt length without ever reaching the desired belt length - Have fun...

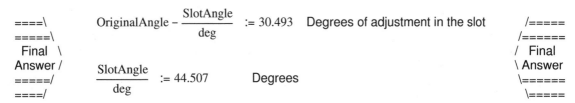

====\
=====\
 Final \
Answer /
=====/
====/

$$\text{OriginalAngle} - \frac{\text{SlotAngle}}{\text{deg}} := 30.493$$ Degrees of adjustment in the slot

$$\frac{\text{SlotAngle}}{\text{deg}} := 44.507$$ Degrees

/=====
/=====
/ Final
\ Answer
\=====
\=====

Exhibit 10-31

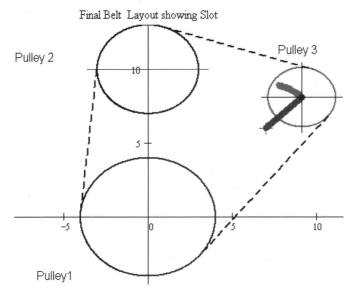

Final Belt Layout showing Slot

Pulley 2

Pulley 3

Pulley1

Exhibit 10-32 Successful solution to the pivot angle.

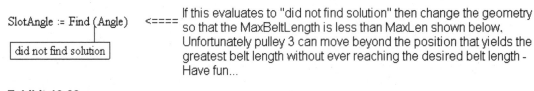

SlotAngle := Find (Angle)   <====

did not find solution

If this evaluates to "did not find solution" then change the geometry so that the MaxBeltLength is less than MaxLen shown below. Unfortunately pulley 3 can move beyond the position that yields the greatest belt length without ever reaching the desired belt length - Have fun...

Exhibit 10-33

the calculation of the nominal belt length and its value. Exhibit 10-30 shows the calculation of the maximum belt length and its value.

Exhibit 10-31 demonstrates how to use the "Solve block" to iterate for a desired solution. The solution is to use the length equation to find the slot angle for that belt length. The angle is set to an initial guess such as the original pivot arm angle less 3°. The word "Given" is entered as the next line. An equation is entered to be solved, followed by another equation that uses the "Find" function. Then the angle of adjustment for the slot is shown in degrees, and then the final slot angle is shown in degrees. The final position of pulley 3 is shown in Exhibit 10-32 along with the adjustment angle for the center of pulley 3.

If the solution cannot be found, then a "did not find solution" message will appear in red as shown in Exhibit 10-33. One problem could be that the "MaxSlotAngle" is too large (it gets

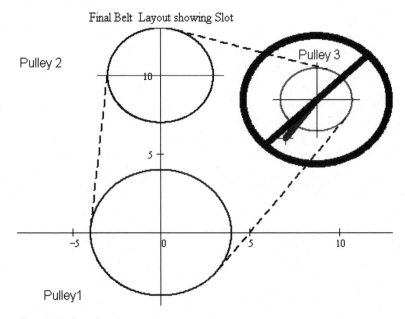

Exhibit 10-34 Unsuccessful pivot angle solution.

NumRows := 13   to display in arrays

$$\text{Angle} := \text{OriginalAngle}, \text{OriginalAngle} - \frac{\text{OriginalAngle} - \text{MaxSlotAngle}}{\text{NumRows}} .. \text{MaxSlotAngle}$$ Range Variable in Degrees

Angle   Length(Angle·deg)

Angle	Length(Angle·deg)
75	49.073
68.462	49.498
61.923	49.889
55.385	50.245
48.846	50.56
42.308	50.832
35.769	51.058
29.231	51.238
22.692	51.368
16.154	51.45
9.615	51.481
3.077	51.461
-3.462	51.392
-10	51.273

Exhibit 10-35

---

smaller as you rotate clockwise) if the belt length monotonically increases. If the MaxSlotAngle cannot be decreased because of physical constraints then proceed with the next adjustments.

Other adjustments require a repositioning of either pulley 3, the pivot arm swing point, or both. These can be determined only by looking at a drawing for the real engine. For our use, the easiest solution is to increase the initial pivot arm angle so that the belt length is shorter and the pivot arm can rotate more degrees before hitting the over-center angle. Normally you would also have to move the pivot arm swing point to the left to accomplish this because pulley 3 is as close to the engine as practical. The position of pulley 3 is shown where the belt reached its maximum length in Exhibit 10-34 along with a "NOT symbol" and the inadequate adjustment angle for the center of pulley 3.

Exhibit 10-35 is frosting on the cake, showing slot angle and belt length arrays. From these arrays it is visually evident that the belt length reached its maximum around 9.6°. The fact that the final slot angle was 44.5° means that the solution is valid. The "NumRows" variable was added to allow easy changes to keep the array from crossing to the next page. Exhibit 10-36 shows how to determine the maximum length of a belt with the current configuration.

$i := 0, 1 .. (\text{OriginalAngle} - \text{MaxSlotAngle}) \cdot 10$ Each 0.1 Degrees to -90 degrees

$$\text{Len}_i := \text{Length}\left[\left(\text{OriginalAngle} - \frac{i}{10}\right) \cdot \text{deg}\right]$$ Length array

$\text{MaxLen} := \max(\text{Len})$

$\text{MaxLen} = 51.481$ <==== Maximum belt length for this pulley
configuration

Exhibit 10-36 Create an array for evaluation.

$\text{Angle} := (\text{OriginalAngle} - 10.) \cdot \text{deg}$ Guess Value

Given

$\text{Length}(\text{Angle}) := \text{MaxLen}$ Equation being solved

$\text{MaxLenAngle} := \text{Find}(\text{Angle})$

$\dfrac{\text{MaxLenAngle}}{\text{deg}} := 8.9$ Degrees for Maximum Length

Exhibit 10-37

on error

Exhibit 10-38 "on error"
statement button.

continue

Exhibit 10-39 "continue"
statement button.

return

Exhibit 10-40 "return"
statement button.

A range variable is set up for each one-tenth of a degree between the original angle and the MaxSlotAngle. This array is then passed to the length function to get an array of belt lengths for each 0.1°. The largest value of the length array is found and displayed. This length must be greater than or equal to the MaxBeltLength to have a valid solution.

Exhibit 10-37 is more cake frosting, as it determines the pivot arm angle where the maximum belt length occurs (over-center angle). This again uses a "Solve block." An initial guess is made of 10° less than the original angle. The word "Given" is entered. In an equation using the "Find" function, belt length is set equal to MaxLen, the pivot arm angle is set equal to the original guess, and this completes the "Solve block." The MaxLenAngle is displayed in degrees for reference.

Additonal buttons used for programming are shown in Exhibits 10-38 to 10-40.

Appendix

When designing accessory mounting brackets for automotive engines weight is a major consideration. This problem is to determine the minimum adjustment angle for an accessory pulley with a belt which has a tolerance of +/- 0.25 inches and a 2% stretch limit.

Pulley 3 is initially positioned as close to the engine as practical and the pivot point is chosen so as to allow the maximum belt stretch to be "taken up" with the adjustment slot for pulley 3. The smaller the slot angle, the less the weight. Once the pivot point and the initial pivot angle are defined, the belt length and slot angle must be found.

Belt Dimensions

$Tolerance := .25$ $BeltStretch := .02$ Percent of Length

Engine Dimensions

$r1 := 4$ Pulley radius $X := 7$ Pivot Location

$r2 := 3$ Pulley radius $Y := 6$ Pivot Location

$r3 := 2$ Pulley radius $L := 3$ Pivot Arm Length

$V := 10$ Vertical distance Pulley 1 to 2 $OriginalAngle := 75$ Pivot Arm Degrees

$MaxSlotAngle := -10.$ Maximum allowable angular position - Degrees

$Angle := OriginalAngle \cdot deg$ radians

$Angle = 1.309$ radians

Exhibit A-1 Belt adjustment program.

$$\text{Wrap1}(A) := \frac{\pi}{2} + \text{atan}\left(\frac{r1 - r2}{V}\right) - \text{atan}\left[\frac{r3 - r1}{\sqrt{(X + L\cdot\cos(A))^2 + (Y + L\cdot\sin(A))^2}}\right] + \text{atan}\left(\frac{Y + L\cdot\sin(A)}{X + L\cdot\cos(A)}\right)$$

$$\text{Len12} := \sqrt{V^2 - (r1 - r2)^2}$$

$$\text{Wrap2}(A) := \frac{\pi}{2} + \text{atan}\left[\frac{r2 - r3}{\sqrt{(X + L\cdot\cos(A))^2 + (V - Y - L\cdot\sin(A))^2}}\right] - \text{atan}\left(\frac{r1 - r2}{V}\right) \ldots$$
$$+ \text{atan}\left(\frac{V - Y - L\cdot\sin(A)}{X + L\cdot\cos(A)}\right)$$

$$\text{Len23}(A) := \sqrt{X^2 + 2\cdot X\cdot L\cdot\cos(A) + (V - Y)^2 - 2\cdot V\cdot L\cdot\sin(A) + 2\cdot Y\cdot L\cdot\sin(A) + L^2 - (r2 - r3)^2}$$

$$\text{Wrap3}(A) := \left[\left[\pi + \text{atan}\left[\frac{r3 - r1}{\sqrt{(X + L\cdot\cos(A))^2 + (Y + L\cdot\sin(A))^2}}\right] \ldots\right. \right. \ldots\right] - \text{atan}\left(\frac{Y + L\cdot\sin(A)}{X + L\cdot\cos(A)}\right)$$
$$+ -\text{atan}\left[\frac{r2 - r3}{\sqrt{(X + L\cdot\cos(A))^2 + (V - Y - L\cdot\sin(A))^2}}\right]$$
$$+ -\text{atan}\left(\frac{V - Y - L\cdot\sin(A)}{X + L\cdot\cos(A)}\right)$$

$$\text{Len31}(A) := \sqrt{X^2 + 2\cdot X\cdot L\cdot\cos(A) + Y^2 + 2\cdot Y\cdot L\cdot\sin(A) + L^2 - (r3 - r1)^2}$$

$$\text{Length}(A) := \text{Wrap1}(A)\cdot r1 + \text{Len12} + \text{Wrap2}(A)\cdot r2 + \text{Len23}(A) + \text{Wrap3}(A)\cdot r3 + \text{Len31}(A)$$

$$\text{MinimumBeltLength} := \text{Length}(\text{Angle}) \qquad \text{Get minimum belt length}$$

$$\text{MinimumBeltLength} = 49.073$$

$$\text{NominalBeltLength} := \text{ceil}\left(\frac{\text{MinimumBeltLength}}{\text{Tolerance}} + 1\right)\cdot\text{Tolerance} \qquad \text{Round length up to tolerance}$$

$$\text{NominalBeltLength} = 49.5$$

$$\text{MaxBeltLength} := (\text{NominalBeltLength} + \text{Tolerance})\cdot(1.0 + \text{BeltStretch}) \qquad \text{Apply tolerance and stretch}$$

$$\text{MaxBeltLength} = 50.745$$

$$\text{Angle} := (\text{OriginalAngle} - 3.)\cdot\text{deg} \qquad \text{Guess Value}$$

Given

$$\text{Length}(\text{Angle}) = \text{MaxBeltLength} \qquad \text{Equation being solved}$$

Exhibit A-1 (*Continued*)

SlotAngle := Find(Angle)   <====   If this evaluates to "did not find solution" then change the geometry so that the MaxBeltLength is less than MaxLen shown below. Unfortunately pulley 3 can move beyond the position that yields the greatest belt length without ever reaching the desired belt length - Have fun...

====\   /=====
=====\   /======
Final \ $\text{OriginalAngle} - \dfrac{\text{SlotAngle}}{\text{deg}} = 30.493$ Degrees of adjustment in the slot   / Final
Answer /   \ Answer
=====/   \======
====/ $\dfrac{\text{SlotAngle}}{\text{deg}} = 44.507$ Degrees   \=====

NumRows := 13   to display in arrays

$\text{Angle} := \text{OriginalAngle}, \text{OriginalAngle} - \dfrac{\text{OriginalAngle} - \text{MaxSlotAngle}}{\text{NumRows}} .. \text{MaxSlotAngle}$ Range Variable in Degrees

Angle	Length(Angle·deg)
75	49.073
68.462	49.498
61.923	49.889
55.385	50.245
48.846	50.56
42.308	50.832
35.769	51.058
29.231	51.238
22.692	51.368
16.154	51.45
9.615	51.481
3.077	51.461
-3.462	51.392
-10	51.273

Create an array for evaluation

$i := 0, 1 .. (\text{OriginalAngle} - \text{MaxSlotAngle}) \cdot 10$

$\text{Len}_i := \text{Length}\left[\left(\text{OriginalAngle} - \dfrac{i}{10}\right) \cdot \text{deg}\right]$ Length array

MaxLen := max(Len)   Maximum belt length for this pulley

MaxLen = 51.481   <====   configuration

Angle := (OriginalAngle − 10.)·deg   Guess Value

Given

Length(Angle)=MaxLen   Equation being solved

MaxLenAngle := Find(Angle)

$\dfrac{\text{MaxLenAngle}}{\text{deg}} = 8.9$ Degrees for Maximum Length

Each 0.1 Degrees to -90 degrees

Exhibit A-1 (*Continued*)

When designing accessory mounting brackets for automotive engines weight is a major consideration. This problem is to determine the minimum adjustment angle for an accessory pulley with a belt which has a tolerance of +/- 0.25 inches and a 2% stretch limit.

Pulley 3 is initially positioned as close to the engine as practical and the pivot point is chosen so as to allow the maximum belt stretch to be "taken up" with the adjustment slot for pulley 3. The smaller the slot angle, the less the weight. Once the pivot point and the initial pivot angle are defined, the belt length and slot angle must be found.

Belt Dimensions

$\text{Tolerance} := .25$ $\text{BeltStretch} := .02$ Percent of Length

Engine Dimensions

$r1 := 4$ Pulley radius $X := 7$ Pivot Location

$r2 := 3$ Pulley radius $Y := 6$ Pivot Location

$r3 := 2$ Pulley radius $L := 3$ Pivot Arm Length

$V := 10$ Vertical distance Pulley 1 to 2 $\text{OriginalAngle} := 75$ Pivot Arm - Degrees

$\text{MaxSlotAngle} := -10.$ Maximum allowable angular position - Degrees

$A := \text{OriginalAngle} \cdot \text{deg}$ radians

$a12 := \text{atan}\left(\dfrac{r1 - r2}{V}\right)$ Compute Belt tangent angle 1 to 2

$a23 := \text{atan}\left[\dfrac{r2 - r3}{\sqrt{(X + L \cdot \cos(A))^2 + (V - Y - L \cdot \sin(A))^2}}\right]$ Compute Belt tangent angle 2 to 3

$a31 := \text{atan}\left[\dfrac{r3 - r1}{\sqrt{(X + L \cdot \cos(A))^2 + (Y + L \cdot \sin(A))^2}}\right]$ Compute Belt tangent angle 3 to 1

$a4 := \text{atan}\left(\dfrac{V - Y - L \cdot \sin(A)}{X + L \cdot \cos(A)}\right)$ Horizontal angle pulleys 2 to 3

$a5 := \text{atan}\left(\dfrac{Y + L \cdot \sin(A)}{X + L \cdot \cos(A)}\right)$ Horizontal angle pulleys 1 to 3

$P0x(\text{ang1}) := L \cdot \cos(\text{ang1}) + X$ Pivot arc (Center of Pulley 3) equations -
showing initial position to maximum slot angle

$P0y(\text{ang1}) := L \cdot \sin(\text{ang1}) + Y$

$\text{ang1} := A, A - .01 .. \text{MaxSlotAngle} \cdot \text{deg}$ range variable

$P1x(\text{ang2}) := r1 \cdot \cos(\text{ang2})$ Pulley 1 circumference equations

$P1y(\text{ang2}) := r1 \cdot \sin(\text{ang2})$

$\text{ang2} := 0, 0.01 .. 2 \cdot \pi$ range variable

$P2x(\text{ang2}) := r2 \cdot \cos(\text{ang2})$

$P2y(\text{ang2}) := r2 \cdot \sin(\text{ang2}) + V$ Pulley 2 circumference equations

Exhibit A-2 Belt adjustment program (with before and after drawings).

$P3y(ang2) := r3 \cdot \sin(ang2) + Y + L \cdot \sin(A)$

$P3x(ang2) := r3 \cdot \cos(ang2) + X + L \cdot \cos(A)$

Pulley 3 circumference equations

$Ymax := V + r2$

$Ymax3 := Y + L \cdot \sin(A) + r3$ Graph Vertical dimensions

$Ymax := if(Ymax > Ymax3, Ymax, Ymax3)$

$Xmax := X + L \cdot \cos(A) + r3$ Graph Horizontal dimension

$Xmax := if(Ymax > Xmax, Ymax, Xmax)$

$x0 := (X - r3 \cdot .2 \quad X + r3 \cdot .2 \quad X \quad X \quad X)^T$ Pivot arm rotation centerlines

$y0 := (Y \quad Y \quad Y \quad Y - r3 \cdot .2 \quad Y + r3 \cdot .2)^T$

$x1 := (-r2 \cdot 1.2 \quad r2 \cdot 1.2)^T$

$y1 := (V \quad V)^T$ Pulley 2 centerline

$j := 0 .. 4$

$k := 0 .. 1$ range variables

$x2 := (X \quad X + L \cdot \cos(A))^T$

$y2 := (Y \quad Y + L \cdot \sin(A))^T$ Pivot Arm

$r7 := r3 \cdot 1.2$ Radius of the NOT sign

$x3 := (X + L \cdot \cos(A) - r7 \quad X + L \cdot \cos(A) + r7 \quad X + L \cdot \cos(A) \quad X + L \cdot \cos(A) \quad X + L \cdot \cos(A))^T$ Pulley 3

$y3 := (Y + L \cdot \sin(A) \quad Y + L \cdot \sin(A) \quad Y + L \cdot \sin(A) \quad Y + L \cdot \sin(A) - r7 \quad Y + L \cdot \sin(A) + r7)^T$ centerlines

$t4 := \pi - a12$

$x4 := (r1 \cdot \cos(t4) \quad r2 \cdot \cos(t4))^T$ Tangent Angle 1 to 2

$y4 := (r1 \cdot \sin(t4) \quad r2 \cdot \sin(t4) + V)^T$ Belt Section 1 to 2

$t5 := \dfrac{\pi}{2} - a23 - a4$ Tangent Angle 2 to 3

$x5 := (r2 \cdot \cos(t5) \quad r3 \cdot \cos(t5) + X + L \cdot \cos(A))^T$ Belt Section 2 to 3

$y5 := (r2 \cdot \sin(t5) + V \quad r3 \cdot \sin(t5) + Y + L \cdot \sin(A))^T$

$t6 := \dfrac{-\pi}{2} - a31 + a5$ Tangent Angle 3 to 1

$x6 := (r3 \cdot \cos(t6) + X + L \cdot \cos(A) \quad r1 \cdot \cos(t6))^T$

$y6 := (r3 \cdot \sin(t6) + Y + L \cdot \sin(A) \quad r1 \cdot \sin(t6))^T$ Belt Section 3 to 1

Exhibit A-2 (*Continued*)

Initial Pulley Layout showing the pivot arm maximum allowable swing motion

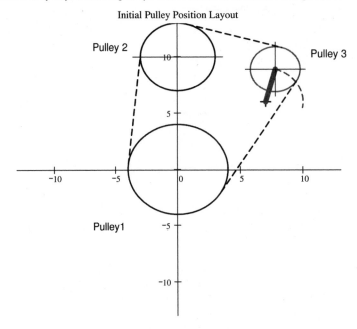

$$\text{Wrap1}(A) := \frac{\pi}{2} + \text{atan}\left(\frac{r1 - r2}{V}\right) - \text{atan}\left[\frac{r3 - r1}{\sqrt{(X + L \cdot \cos(A))^2 + (Y + L \cdot \sin(A))^2}}\right] + \text{atan}\left(\frac{Y + L \cdot \sin(A)}{X + L \cdot \cos(A)}\right)$$

$$\text{Len12} := \sqrt{V^2 - (r1 - r2)^2}$$

$$\text{Wrap2}(A) := \frac{\pi}{2} + \text{atan}\left[\frac{r2 - r3}{\sqrt{(X + L \cdot \cos(A))^2 + (V - Y - L \cdot \sin(A))^2}}\right] - \text{atan}\left(\frac{r1 - r2}{V}\right) \dots$$
$$+ \text{atan}\left(\frac{V - Y - L \cdot \sin(A)}{X + L \cdot \cos(A)}\right)$$

$$\text{Len23}(A) := \sqrt{X^2 + 2 \cdot X \cdot L \cdot \cos(A) + (V - Y)^2 - 2 \cdot V \cdot L \cdot \sin(A) + 2 \cdot Y \cdot L \cdot \sin(A) + L^2 - (r2 - r3)^2}$$

$$\text{Wrap3}(A) := \left[\left[\pi + \text{atan}\left[\frac{r3 - r1}{\sqrt{(X + L \cdot \cos(A))^2 + (Y + L \cdot \sin(A))^2}}\right] \dots \right] \dots \right] - \text{atan}\left(\frac{Y + L \cdot \sin(A)}{X + L \cdot \cos(A)}\right)$$
$$+ -\text{atan}\left[\frac{r2 - r3}{\sqrt{(X + L \cdot \cos(A))^2 + (V - Y - L \cdot \sin(A))^2}}\right]$$
$$+ -\text{atan}\left(\frac{V - Y - L \cdot \sin(A)}{X + L \cdot \cos(A)}\right)$$

$$\text{Len31}(A) := \sqrt{X^2 + 2 \cdot X \cdot L \cdot \cos(A) + Y^2 + 2 \cdot Y \cdot L \cdot \sin(A) + L^2 - (r3 - r1)^2}$$

Exhibit A-2 (*Continued*)

$\text{Length}(A) := \text{Wrap1}(A) \cdot r1 + \text{Len12} + \text{Wrap2}(A) \cdot r2 + \text{Len23}(A) + \text{Wrap3}(A) \cdot r3 + \text{Len31}(A)$

$\text{Angle} := \text{OriginalAngle} \cdot \text{deg}$

$\text{MaxBeltLen} := \text{Tolerance} \cdot \left(\text{ceil} \left(\dfrac{\overset{\text{radians}}{\text{Length}(\text{Angle}) + \text{Tolerance}}}{\text{Tolerance}} \right) + 1. \right) \cdot (1. + \text{BeltStretch})$

$\text{SL}(PA) := \text{Length}(PA) - \text{MaxBeltLen}$ Equation to be solved

$\text{Angle} := (\text{OriginalAngle} - 3.) \cdot \text{deg}$ Guess Value

Given

$\text{SL}(\text{Angle}) = 0$ Equation being solved

$\text{SlotAngle} := \text{Find}(\text{Angle})$ <==== If this evaluates to "did not find solution" then change the geometry. Unfortunately pulley 3 can move beyond the position that yields the greatest belt length without ever reaching the desired belt length - Have fun...

```
====\
=====\
Final  \     OriginalAngle - SlotAngle/deg = 30.493      Degrees of adjustment in the slot
Answer /
=====/     SlotAngle/deg = 44.507      Degrees
====/
```

$\text{OriginalAngle} - \dfrac{\text{SlotAngle}}{\text{deg}} = 30.493$ Degrees of adjustment in the slot

$\dfrac{\text{SlotAngle}}{\text{deg}} = 44.507$ Degrees

```
/=====
/======
/ Final
\ Answer
\======
\=====
```

$\text{MaxBeltLen} = 50.745$

$i := 0, 1 \, .. \, (\text{OriginalAngle} - \text{MaxSlotAngle}) \cdot 10$ Each 0.1 Degrees to -90 degrees

$\text{Len}_i := \text{Length}\left[\left(\text{OriginalAngle} - \dfrac{i}{10} \right) \cdot \text{deg} \right]$ Length array containing maximum

$\text{MaxLen} := \max(\text{Len})$

$\text{MaxLen} = 51.481$ <==== Maximum belt length for this pulley configuration

$A := \text{SlotAngle}$

$\text{ang3} := \text{OriginalAngle} \cdot \text{deg}, \text{OriginalAngle} \cdot \text{deg} - .01 \, .. \, A$ range variable

$\text{P3y}(\text{ang2}) := r3 \cdot \sin(\text{ang2}) + Y + L \cdot \sin(A)$

$\text{P3x}(\text{ang2}) := r3 \cdot \cos(\text{ang2}) + X + L \cdot \cos(A)$ Pulley 3 circumference equations

$\text{Ymax3} := Y + L \cdot \sin(A) + r3$

$\text{Ymax} := \text{if}(\text{Xmax} > \text{Ymax3}, \text{Xmax}, \text{Ymax3})$ Graph Vertical dimensions

$\text{Xmax} := X + L \cdot \cos(A) + r3$

$\text{Xmax} := \text{if}(\text{Ymax} > \text{Xmax}, \text{Ymax}, \text{Xmax})$ Graph Horizontal dimension

$\text{x2} := (X \quad X + L \cdot \cos(A))^T$

$\text{y2} := (Y \quad Y + L \cdot \sin(A))^T$ Pivot Arm

$\text{x3} := (X + L \cdot \cos(A) - r7 \quad X + L \cdot \cos(A) + r7 \quad X + L \cdot \cos(A) \quad X + L \cdot \cos(A) \quad X + L \cdot \cos(A))^T$

$\text{y3} := (Y + L \cdot \sin(A) \quad Y + L \cdot \sin(A) \quad Y + L \cdot \sin(A) \quad Y + L \cdot \sin(A) - r7 \quad Y + L \cdot \sin(A) + r7)^T$ Pulley 3 centerlines

Exhibit A-2 (*Continued*)

$$t5 := \frac{\pi}{2} - a23 - atan\left(\frac{V - Y - L \cdot sin(A)}{X + L \cdot cos(A)}\right)$$ Tangent Angle 2 to 3

$$x5 := (r2 \cdot cos(t5) \quad r3 \cdot cos(t5) + X + L \cdot cos(A))^T$$

$$y5 := (r2 \cdot sin(t5) + V \quad r3 \cdot sin(t5) + Y + L \cdot sin(A))^T$$ Belt Section 2 to 3

$$t6 := \frac{-\pi}{2} - a31 + atan\left(\frac{Y + L \cdot sin(A)}{X + L \cdot cos(A)}\right)$$ Tangent Angle 3 to 1

$$x6 := (r3 \cdot cos(t6) + X + L \cdot cos(A) \quad r1 \cdot cos(t6))^T$$

$$y6 := (r3 \cdot sin(t6) + Y + L \cdot sin(A) \quad r1 \cdot sin(t6))^T$$ Belt Section 3 to 1

$$r7 := if(MaxBeltLen > MaxLen, r7 \cdot 1.2, 0)$$ Not" sign radius

$$x7 := (X + L \cdot cos(A) - r7 \quad X + L \cdot cos(A) + r7)^T$$

$$y7 := (Y + L \cdot sin(A) - r7 \quad Y + L \cdot sin(A) + r7)^T$$ Not" sign slash

$$P7y(ang2) := r7 \cdot sin(ang2) \cdot 1.414 + Y + L \cdot sin(A)$$

$$P7x(ang2) := r7 \cdot cos(ang2) \cdot 1.414 + X + L \cdot cos(A)$$ Not" sign circle

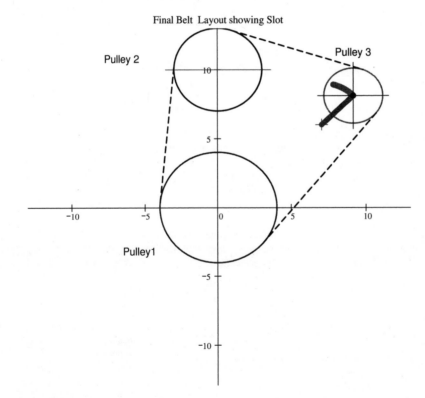

Final Belt Layout showing Slot

Exhibit A-2 (*Continued*)

Index

ABOUT THE AUTHORS

HOWARD KELLER recently retired from the Corporate Planning Department of MetLife. Previously he was a physicist at the Buhl Planetarium and Institute of Popular Science in Pittsburgh, as well as an industrial engineer in the steel industry. Mr. Keller is co-author of *The Right Mix,* also published by McGraw-Hill. JOHN G. CRANDALL has worked as a mechanical engineer at General Motors, and is a founder of Duquesne Systems and a former vice president of information systems at Legent. His latest venture is Optimum Power Technology, which specializes in engine simulation software.